T0155959

Lecture Notes in Computer Science 13425

More information about this series at https://link.springer.com/bookseries/558

Karol Desnos · Sergio Pertuz (Eds.)

Design and Architecture for Signal and Image Processing

15th International Workshop, DASIP 2022
Budapest, Hungary, June 20–22, 2022
Proceedings

Editors
Karol Desnos
IETR, INSA
Rennes, France

Sergio Pertuz
TU Dresden
Dresden, Germany

ISSN 0302-9743 ISSN 1611-3349 (electronic)
Lecture Notes in Computer Science
ISBN 978-3-031-12747-2 ISBN 978-3-031-12748-9 (eBook)
https://doi.org/10.1007/978-3-031-12748-9

This Springer imprint is published by the registered company Springer Nature Switzerland AG
The registered company address is: Gewerbestrasse 11, 6330 Cham, Switzerland

Preface

This volume contains papers presented at the 15th Workshop on Design and Architectures for Signal and Image Processing (DASIP 2022), which was held jointly with the 17th HiPEAC Conference, during June 20–22, 2022, in Budapest, Hungary. DASIP provides an inspiring international forum for the latest innovations and developments in the field of leading signal, image, and video processing and machine learning in custom embedded, edge, and cloud computing architectures and systems.

This year, two calls for papers were organized, the first in fall 2021 and the second in spring 2022. We received 32 paper submissions from authors in 11 countries around the world, and 13 high-quality papers were accepted as oral presentations. Each contributed paper underwent a rigorous double-blind peer review process during which it was reviewed by at least three reviewers who were drawn from a large pool of the Technical Program Committee members.

The success of DASIP 2022 depended on the contributions of many individuals and organizations. With that in mind, we thank all authors who submitted their work to the conference. We also wish to offer our sincere thanks to the members of the Technical Program Committee, for their very detailed reviews, and to the members of the Steering Committee.

We would also like to address special thanks to Dave Lacey, from Graphcore (UK), and Eduardo Juarez, from the Universidad Politecnica de Madrid (Spain), for presenting deeply inspiring keynotes during the event.

June 2022

Karol Desnos
Sergio Pertuz

Organization

General Chairs

Karol Desnos	IETR, INSA Rennes, France
Sergio Pertuz	Technische Universität Dresden, Germany

Steering Committee

Bertrand Granado	Sorbonne University, France
Diana Goehringer	Technical University of Dresden, Germany
Eduardo de la Torre	Polytechnic University of Madrid, Spain
Guy Gogniat	University of Southern Brittany, France
Jean-Francois Nezan	IETR, INSA Rennes, France
Jean-Pierre David	Polytechnique Montréal, Canada
Joao M. P. Cardoso	University of Porto, Portugal
Marek Gorgon	AGH University of Science and Technology, Poland
Michael Huebner	Brandenburg University of Technology, Germany
Paolo Meloni	University of Cagliari, Italy
Pierre Langlois	Polytechnique Montréal, Canada
Sebastien Pillement	University of Nantes, France
Tomasz Kryjak	AGH University of Science and Technology, Poland

Program Committee

Francois Berry	Institut Pascal, CNRS, University of Clermont-Auvergne, France
Arnaud Bourge	STMicroelectronics, France
Jani Boutellier	University of Vaasa, Finland
Gabriel Caffarena	University CEU San Pablo, Spain
Juan Carlos Lopez	University of Castilla-La Mancha, Spain
Daniel Chillet	IRISA/ENSSAT, University of Rennes 1, France
Martin Danek	daiteq s.r.o., Czech Republic
Milos Drutarovsky	Technical University of Kosice, Slovakia
Joao Canas Ferreira	University of Porto, Portugal
Oscar Gustafsson	Linkoping University, Sweden
Frank Hannig	University of Erlangen-Nurnberg, Gemany

Dominique Houzet	Grenoble Institute of Technology, France
Mateusz Komorkiewicz	Aptive, Poland
Lionel Lacassagne	Sorbonne University, France
Ahmed Lakhssassi	Universite du Quebec en Outaouais, Canada
Yannick Le Moullec	Tallinn University of Technology, Estonia
Johan Lilius	Abo Akademi University, Finland
Sebastian Lopez	University of Las Palmas de Gran Canaria, Spain
Gustavo Marrero Callico	University of Las Palmas de Gran Canaria, Spain
Kevin J. M. Martin	University of Southern Brittany, France
Gabriela Nicolescu	Polytechnique Montréal, Canada
Jari Nurmi	Tampere University, Finland
Arnaldo Oliveira	University of Aveiro, Portugal
Andres Otero	Polytechnic University of Madrid, Spain
Francesca Palumbo	University of Sassari, Italy
Maxime Pelcat	IETR, INSA Rennes, France
Fernando Pescador	Polytechnic University of Madrid, Spain
Christian Pilato	Polytechnic University of Milan, Italy
Andrea Pinna	Sorbonne University, France
Jorge Portilla	Polytechnic University of Madrid, Spain
Alfonso Rodriguez	Polytechnic University of Madrid, Spain
Nuno Roma	University of Lisbon, Portugal
Olivier Romain	CY Cergy Paris University, France
Paweł Russek	AGH University of Science and Technology, Poland
Ruben Salvador	CentraleSupélec, France
Carlo Sau	University of Cagliari, Italy
Yves Sorel	Inria, France
Dimitrios Soudris	National Technical University of Athens, Greece
Walter Stechele	Technical University of Munich, Germany
Marcin Szelest	Aptive, Poland
Claude Thibeault	Ecole de Technologie Superieure, Canada
Jose Vieira	University of Aveiro, Portugal
Tanya Vladimirova	University of Leicester, UK
Serge Weber	University of Lorraine, France

Additional Reviewers

Ivan Luca Costa
Nuno Neves
João Vieira

Contents

Software and Architecture for Telecommunication Systems

High-Performance Gallager-E Decoders for Hard Input LDPC Decoding on Multi-core Devices

Bertrand Le Gal(✉), Vincent Pignoly, and Christophe Jego

IMS laboratory (UMR 5218), Bordeaux INP, Univ. Bordeaux, Talence, France
{bertrand.legal,vincent.pignoly,christophe.jego}@ims-bordeaux.fr

Abstract. LDPC codes are a family of error-correcting codes that are present in most space communication standards. Thanks to their large processing power and their parallelization capabilities, prevailing multi-core devices facilitate real-time implementations of digital communication systems, which were previously implemented thanks to dedicated hardware circuits. A lot of works were done over the last decade on the implementation of Gbps decoders on programmable devices. However, these works focus on soft-input LDPC decoding algorithms. But, hard-input LDPC decoders are also required to design and prototype optical-based satellite communication systems. In this article, the first software based implementation of a hard-input multi-Gbps LDPC decoder is detailed. Thanks to different parallelization strategies and deeply optimized SIMD codes, throughputs up to 7.5 Gbps are achieved when 10 Gallager-E iterations are executed onto an INTEL Xeon device.

Keywords: LDPC · Gallager E · multi-core · SIMD · High-throughput

1 Introduction

Low-Density Parity-Check (LDPC) codes are a popular class of Error Correction Codes (ECC) used in digital communication systems to provide reliability. Due to their excellent error correction performance, LDPC codes were selected for terrestrial wireless standards (e.g., WiFi and 5G) but also in RF space ones (CCSDS, DVB-S2 and DVB-S2x). FPGA or ASIC technologies were during a long time the single way to provide real-time LDPC decoding when hundreds of Mbps or Gpbs are targeted. Indeed, LDPC decoding algorithms are characterized by a high computational complexity discarding other kinds of implementations. These dedicated hardware implementations provide high-throughput and low-energy features at the costs of low-flexibility and low-reusability.

For ten years, the computational power offered by multi-core or many-core devices associated with easy-to-use programming models opened new horizons. Indeed, coping with low-flexibility and long development cycles, researchers and industrials tried to use these programmable devices to implement ECC decoders

K. Desnos and S. Pertuz (Eds.): DASIP 2022, LNCS 13425, pp. 3–15, 2022.
https://doi.org/10.1007/978-3-031-12748-9_1

that are the receiver design bottlenecks [1,2,11,18]. Programmable architectures associated with optimized software descriptions made possible the implementation of high-throughput receiver systems. They can be used as real life wireless communication systems and/or prototype for next generation ones.

Software Defined Radio (SDR) [9] or cloud-RAN [3,21] systems were targeted by previous works. In these works, the RF front end and demodulation stage provide soft-input to the LDPC decoding process. Consequently, like in the field of ASIC/FPGA LDPC decoders, previous works focused on soft-input Min-Sum (MS) algorithm. Unfortunately, to reach a high throughput of several Gbps in optical space communications, ECC decoders can be limited to process hard input values due to current optical technology limitations.

Hard-input decoding presents lower error correction performance than soft-input ones. Moreover, it needs the implementation of other LDPC decoding algorithms such as Gallager-B, Gallager-E or their variations (e.g., Bit flipping algorithms [10], Probabilistic Gallager-B [19]). Even if efficient FPGA implementations of Probabilistic Gallager B (PGaB) and Gallager-E are detailed in [16,19]. Implementing them efficiently in software is not an easy task. From a software point of view, the PGaB decoding algorithm needs random number generation and involves computation hazards at runtime making it clearly incompatible with processor features and thus inappropriate for Gbps performance. At the opposite, the Gallager-E decoding algorithm has a formulation closed to the MS one used in related works. Its computation parallelism is almost regular but its high memory footprint and the logical bit-level computations are not clearly adapted to software processor targets. However, in this work we focused on it and propose its efficient SIMD (Single Instruction Multiple Data) and SIMT (Single Instruction Multiple Threads) implementation.

The remainder of the paper is organized as follows. Section 2 introduces LDPC codes and the horizontal-layered Gallager-E decoding algorithm. Then, the parallelization strategy and the applied optimizations are provided in Sect. 3. Section 4 summarizes the experimental results obtained with the proposed decoder implementations. Finally, conclusion and future works are reported.

2 LDPC Decoding Algorithm

An LDPC code is a linear block code defined by a binary sparse $M \times N$ parity-check matrix called $\mathbf{H}$. This $\mathbf{H}$ matrix is composed of N columns representing the received bit information (VN) from the channel whereas the $M = N - K$ rows are associated to parity-check information (CN) with K the number of information bits in the received frame. To ease the implementation of LDPC decoders, a special class of LDPC codes is used in standards. Quasi-Cyclic (QC) LDPC codes are codes that are composed of an array of $Z \times Z$ shifted identity sub-matrices. Z is the order that defines the computation parallelism level. It eases $\mathbf{H}$ matrix storage and ensures that Z CNs can be processed in parallel without conflicts independently of the computation scheduling.

Algorithm 1. Horizontal-layered Gallager E

▷ Input (received word) : $\boldsymbol{y} = (y_1, y_1, ..., y_N) \in \{0,1\}^N$
init
  $t = 0$, $Y_i = 2y_i - 1$ and $V_i = 0$ for $i \in [1, .., N]$
repeat
  ▷ **L1** - loop 1: Horizontal layered scheduling
  for all j $\in [1, .., M]$ **do**
    $C_j^{(t)} = 1$, $Sum0 = 0$
    ▷ **L2** - Loop 2
    for all i $\in$ N(j) **do**
      ▷ **L1S1** - Stage S1: Variable to parity check message $v2c_{ij}^{(t)}$ processing

$$M_{ij} = \begin{cases} V_i, & t = 0 \\ V_i - c2v_{ji}^{(t-1)}, & \text{otherwise} \end{cases}$$

$$v2c_{ij}^{(t)} = sign(M_{ij} + \omega^{(t)} * Y_i)$$

      ▷ **L1S2** - Stage S2: Check node $C_j^{(t)}$ processing

$$C_j^{(t)} = \begin{cases} C_j^{(t)}, & M_{ij} \geqslant 0 \\ -C_j^{(t)}, & \text{otherwise} \end{cases}$$

      if $M_{ij} = 0$ **then** $Sum0 = Sum0 + 1$ **end if**
    end for
    ▷ **L3** - Loop 3
    for all i $\in$ N(j) **do**
      ▷ **L3S1** - Stage S1: Check to variable message $c2v_{ji}^{(t+1)}$ processing

$$c2v_{ji}^{(t)} = \begin{cases} 0, & Sum0 \geqslant 2 \\ 0, & Sum0 = 1 \text{ and } M_{ij}^{(t)} \neq 0 \\ C_j^{(t)}, & Sum0 = 1 \text{ and } M_{ij}^{(t)} = 0 \\ C_j^{(t)} \times M_{ij}^{(t)}, & \text{otherwise} \end{cases}$$

      ▷ **L3S2** - Stage S2: Variable node accumulator V_i updating

$$V_i = M_{ij} + c2v_{ji}^{(t+1)}$$

    end for
  end for
  $t = t + 1$
until $t \leq t_{max}$

$$\forall i \in [1, .., N],\ x_i = \begin{cases} y_i, & Y_i + V_i = 0 \\ 0, & Y_i + V_i > 0 \\ 1, & Y_i + V_i < 0 \end{cases}$$

▷ Output (decoded word) : $\boldsymbol{x} = (x_1, x_1, ..., x_N) \in \{0,1\}^N$

Usually, the LDPC decoding process is performed thanks to a message passing (MP) approach where VNs and CNs exchange m messages. When the decoding process benefits from soft-input, the sum-product algorithm (SPA) approximations such as Min-Sum (MS) variants are applied [4,13].

Hard-input constraint involves another algorithmic choice. Gallager-B algorithm was initially proposed in [5] to process hard-input values. However, its decoding performance is relatively low due to binary values used to represent exchanged messages. This algorithm was recently improved in terms of error correction power by inserting for instance decoding noise (e.g. Probabilistic Gallager-B [19]) or using gradient descent based algorithm [7,20]. These algorithms (PGaB, GDBF and PGDBF) were developed for hardware decoders manipulating for short frame sizes. However, in the current context, they are useless due to spatial related constraints: (a) the codewords should be long (> 16k bits), and (b) to reach high-throughput performances randomness and computation irregularity should be avoided. Consequently, they were discarded. Original Gallager B decoding algorithm [14] can be extended by considering erasures in message passing. This extended algorithm, called Gallager E algorithm in [17], manages ternary values $-1, 0, +1$ for exchanged messages instead of binary $-1, +1$ ones in the Gallager B algorithm. This third value that represents doubt during the decoding process, drastically improves error correction performance.

Gallager E algorithm is summarized in Algorithm 1. Horizontal layered scheduling was selected because it improves error correction performances whereas at the same time it reduces computation and memory complexities [16]. Algorithmic structure is closed to the one used for MS decoding [15]. However, contrary to MS algorithms that execute 8b arithmetic operations (addition, minimum and comparison), Gallager E algorithm needs on its side 1b or 2b logic operations and conditional branches to execute non-reversible operations such as voting.

Received bits Y_i come from the channel. The initial values of accumulator nodes V_0 should be null as presented in [16]. $C^{(tj}$ corresponds to the check node values during the iteration t, with t the current iteration and t_{max} the maximum number of decoding iterations. $V_{ij}^{(t)}$ represents the value of the vote for node VN_i. The possible value set for $V_{ij}^{(t)}$ message is in range $\{-1, 0, +1\}$. Message $c2v_{ji}^{(t+1)}$ from CN_j to VN_i is the product of incoming messages except $v2c_{ij}^{(t)}$. If one or more input messages equal zero then the output message equals the null value too. In this step, contrary to MS decoding algorithm, the $c2v_{ji}^{(t+1)}$ value cannot be easily deduced from $v2c_{ij}^{(t)}$ because of the vote operation that cannot be inverted. Moreover, as the vote operation could not be inverted, the Y_i values should be kept in memory during the overall decoding process. It increases the memory footprint of N elements in comparison to MS decoder implementations.

3 Parallelization Strategies

3.1 Targeted Multi-core System

INTEL x86 processors currently provide different parallelization features. At the higher level, the processor circuits include many independent physical cores that can be used to process tasks or sub-tasks in parallel (SIMT). The number of cores can reach up to dozens for server grade processors. At the same time, each physical processor core includes SIMD (Single Instruction Multiple Data) units that execute parallel computations. SIMD units are 128b up to 512b wide, authorizing up to 64× 8b computations per instruction. Finally, at the lower level, x86 processors are superscalar and thus implements Instruction-Level Parallelism (ILP) authorizing multiple instructions to be scheduled within a single clock cycle according to the resource availability. To reach efficiency, all these aspects should be addressed together. The LDPC decoding parallelization strategy and the optimizations are reported in this section.

3.2 SIMD Parallelization

Parallelization of message passing algorithm was widely studied in the context of traditional MS decoder implementations [1]. Two main SIMD parallelization strategies for multi-core targets where proposed [2,11]. These generic approaches that can also be applied for the Gallager-E algorithm provide different advantages and drawback effects.

Inter-frame parallelization strategy [2] takes advantage of SIMD units to decode multiple frames in parallel. It eases the software description and provides regular computation parallelism at runtime. Indeed, when the number of frames processed in parallel ($Q \times 8$b) equals the SIMD width, the SIMD efficiency is constant to 100%. The inter-frame strategy drawback effect comes from the memory footprint (Δ_{inter}). Indeed, this footprint becomes quickly larger than L1, L2 and L3 memory caches. Its high memory bandwidth requirement limits the scalability of the decoder implementations but also provides high processing latency and impact on global system performance due.

$$\Delta_{\text{inter}} = Q \times (2 \times N + m) + m \tag{1}$$

At the opposite, intra-frame strategy [11] takes advantage of SIMD units to parallelize internal computations from a single frame. This strategy limits the memory footprint (Δ_{intra}) at runtime. However, it complexifies the software description of the decoder and slow down memory accesses. Indeed, **H** structure management is done at runtime and involves complex memory accesses. Moreover, depending on the LDPC code, a SIMD usage rate of 100% is rarely obtained. However, from a system point of view, the intra-frame implementation delivers low-latency feature and limits its impact on other processing elements.

$$\Delta_{\text{intra}} = 2 \times N + m + \frac{m}{Z} \tag{2}$$

Both SIMD parallelization strategies can be applied to speed-up the execution of loop 1 defined in Algorithm 1. In this work, both approaches are evaluated because they provide different features and thus different trade-offs.

3.3 ILP Improvement

An efficient implementation of loops 2 and 3 in Algorithm 1 is crucial because they are executed M times per decoding iteration. Consequently, a specific mapping of the algorithmic operations on available SIMD instruction is required. Gallager-E decoding algorithm mainly manipulates bit or ternary values and has many conditional instructions. So Gallager-E decoding is more challenging to efficiently implement on processor cores than MS ones [15].

First, to reduce the complexity of the M_{ij} computation (**L2S1**) that depends on the decoding iterations, an initialization of the messages to zero was done before the decoding. It avoids the comparison and conditional moves at runtime. Then for the computation of $v2c$ messages, as ω is in range 0, 1, the multiplication operation is implemented thanks to a logical *and* instruction whose second operand is a binary mask (0x00, 0xFF). Finally, the counter (Sum0) can be approximated: its value is increased by the result of the comparison instruction that is $\{0x00, 0xFF\}$ and nor by 1. This approximation is possible because in **L3S1**, the first part of the conditional structure can be reformulated as $Sum0 - (M_{ij} = 0) > 0$. This tricky optimization removes logical instructions and comparisons from the execution critical path. Moreover, it eases the implementation of the $c2v$ conditional computation making possible to describe it as a value selection in range $0, 1$, and then a conditional sign inversion to regenerate a message in range $\{-1, 0, +1\}$.

After these transformations, the number of instructions needed in the processing kernels (loops **L2** and **L3**) is quite small whereas the number of **L2**/**L3** loop iterations is limited to range $[7, 20]$ (due to benchmarked LDPC codes). Consequently, to improve the ILP thanks to instruction execution overlapping and also remove useless control instructions, specialized kernel codes are generated at compile time. To this end, the features of C++11 language (i.e., template specialization) are applied like in [8]. For each CN degree value, a dedicated and optimized kernel is generated. Consequently, the number of instructions is then minimized for each **L1** loop execution. At runtime, an array of function pointers is used to select the right binary code to execute.

3.4 Memory Compression

The memory bandwidth is a bottleneck for software implementation of LDPC decoders when long frame are processed. Decoder memory footprint becomes quickly higher than memory caches. Contrary to MS-based decoder implementations that manage 8b values internally for all data, Gallager-E decoding algorithm manipulate only bit or ternary values. Naively, all these values consume 8b in memory because they are involved in 8b arithmetic operations. However, as the memory bandwidth is a bottleneck, a memory compression technique was developed. It divides the memory footprint by 4 for exchanged messages ($c2v$) that are ternary values and divides by 8 the footprint for channel values (Y_i) that are binary values. Memory footprint reduction necessitates the execution of additional SIMD instructions at runtime. However, both compression and decompression tasks are executed with a low latency penalty on x86 architecture whereas a single memory cache miss can produce a penalty of some hundreds of clock cycles. The compression and the decompression tasks can be done easily and efficiently thanks to the source codes provided in Listing 1. Note that data compression does not impact on error correction performances because values that are stored on 8b are in reality 2b or 1b.

3.5 SPMD Parallelization

Different parallelization techniques could be applied to take advantage of the **P** cores for message passing algorithm implementation. It is possible to use them to speed-up the execution of the loop 1 (**L1**) defined in Algorithm 1. However, this

```
const __m512i zero    = _mm512_setzero_si512();
const __m512i pos_one = _mm512_set1_epi8(0x01);
const __m512i neg_one = _mm512_set1_epi8(0xFF);

void compress_and_store_yi(__mmask64* ptr, const __m512i x) {
    ptr[0] = _mm512_movepi8_mask(x);
}

__m512i load_and_uncompress_yi(const __mmask64 x) {
    return _mm512_mask_blend_epi8( x, r_one, neg_one );
}

void compress_and_store_msg(__mmask64* ptr, const __m512i x) {
    ptr[0] = _mm512_cmpeq_epi8_mask( x, pos_one );
    ptr[1] = _mm512_cmpeq_epi8_mask( x, neg_one );
}

__m512i load_and_uncompress_msg(const __mmask64* ptr) {
    __m512i w1 = _mm512_mask_blend_epi8( ptr[0], zero, pos_one );
    __m512i w2 = _mm512_mask_blend_epi8( ptr[1], zero, neg_one );
    return _mm512_or_si512(w1, w2);
}
```

Listing 1.1. SIMD functions used for (de)compression of binary and ternary values

approach is inefficient because: (1) loop elements are not independent when horizontal layered based decoding algorithm is applied due to the fact that memory access conflicts can occur, and (2) the time spend in forking and joining tasks is not negligible compared to **L1** execution time. The best way to increase the performance level is to allocate **P** independent LDPC decoders to process **P** distinct frames. It avoids L1/L2 memory sharing at runtime between the cores but increase L3 cache usage. It also increases the pressure on the memory bandwidth by a factor **P**. In the current work to enable an asynchronous behavior of the decoders, the LDPC decoders are encapsulated in C++11 threads.

4 Experimentation Results

4.1 Experimentation Setup

The software-based LDPC decoder implementations were developed in C++ 11 language. The targeted device was an INTEL Xeon Gold processor. Consequently, the AVX512 instruction subset was selected. To benefits from INTEL SIMD features, INTEL intrinsics which are C-style functions were applied. To optimize the instruction scheduling and generate the executable file, the software decoder descriptions were compiled with the CLANG++/LLVM 10.0 toolchain. The compilation flags provided to the toolchain are: *-march=native -mtune=native -Ofast -funroll-loops*.

The host platform was a multi-core system composed of a dual socket INTEL Xeon Gold 6148 CPU. Each Xeon processor has 20 physical processor cores. The overall processor cores shares a 28160K L3 memory cache and 256 GB of RAM. A working frequency up to 3, 70 GHz is achievable on this platform thanks

Table 1. Selected QC-LDPC codes for benchmarking purposes

Code	C_1 [6]	C_2	C_3
(N, K)	(1296, 648)	(32768, 16384)	(20480, 16384)
Z	54	256	256
d_c	{8}	{7, 8, 9}	{19, 20}
# msgs	5184	131072	81664

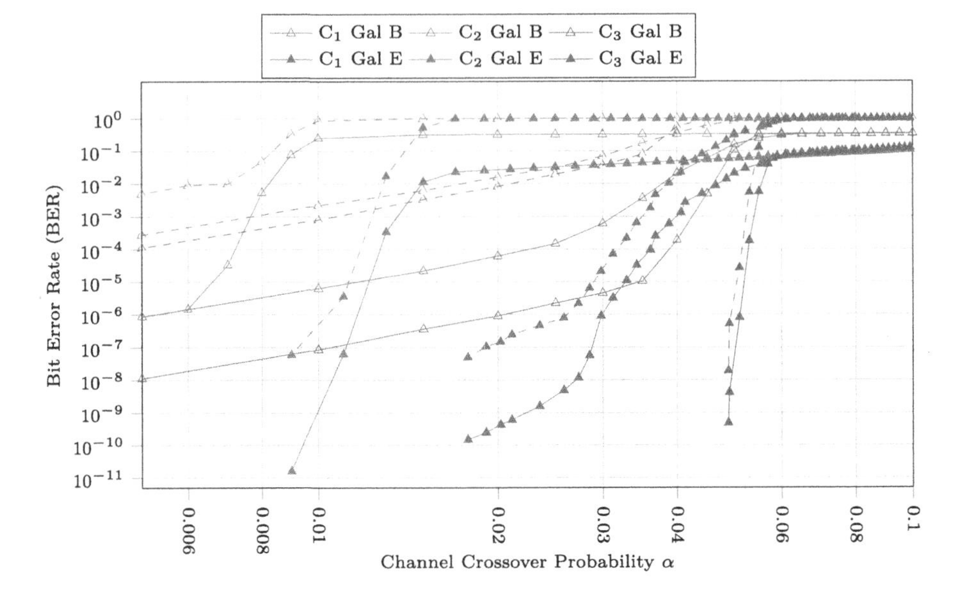

Fig. 1. Error correction performance comparison of Gallager-B and Gallager-E algorithms when 10 layered decoding iterations are executed.

to turbo boost feature when a single processor core is activated. The average working frequency is 2,40 GHz and 2,20 GHz when 50% and 100% of the cores are activated, respectively. This frequency reduction is due to power dissipation constraints. Note that these values were confirmed using the *i7z* tool during experiments.

4.2 Error Correction Performance

Before benchmarking the throughput efficiency of the proposed Gallager-E parallelization schemes, a validation of its BER and FER performances was done. To check its behavior, three different QC-LDPC codes were used. The first code comes from related works on hard-input LDPC decoding [6]. The two others are custom LDPC codes developed specifically for spatial optical communications [16]. The main characteristics of the codes are summarized in Table 1.

Table 2. Performances of Gallager-E LDPC decoders on INTEL Xeon Gold 6148 CPU

		Γ in Mbit/s			Δ in μs			**P** in Watts			**e** in nJ/bit		
Code	#cores	Γ_{d_1}	Γ_{d_2}	Γ_{d_3}	Δ_{d_1}	Δ_{d_2}	Δ_{d_3}	$\mathbf{P}_{d_1}$	$\mathbf{P}_{d_2}$	$\mathbf{P}_{d_3}$	$\mathbf{e}_{d_1}$	$\mathbf{e}_{d_2}$	$\mathbf{e}_{d_3}$
C_1	1	294	272	220	281	304	6	180	180	180	613	662	819
C_2	1	108	136	180	12034	9523	113	167	167	172	1547	1228	956
C_3	1	77	106	247	27010	19671	132	171	170	169	2221	1228	685
C_1	20	4487	4041	3207	369	410	8	300	299	290	67	74	91
C_2	20	813	2412	2546	53106	10865	162	419	411	297	516	171	117
C_3	20	876	2030	3177	48785	21192	218	412	416	298	471	205	94
C_1	40	7532	6833	5460	440	485	10	351	342	331	47	51	61
C_2	40	784	3447	4298	66739	15213	191	437	472	340	558	137	80
C_3	40	712	2286	5446	118795	37893	248	434	473	341	610	207	63

The bit error rate (BER) and frame error rate (FER) performances for C_1 to C_3 LDPC codes are reported in Fig. 1 when a Monte-Carlo simulation is built on a BSC channel and a OOK modulation. As expected, the curves show that Gallager-E algorithm outperforms the Gallager-B ones in the overall use-cases. These results are consistent with the published literature. They demonstrate the correct functionality of the LDPC decoders implemented in this work.

4.3 Absolute Performances

For benchmarking purpose, a communication system simulation runs during a period of 120 seconds to avoid working frequency scaling impact on average values. The throughput (Γ) and the latency (Δ) results are reported in Table 2 when 10 decoding iterations are executed. The following setups are evaluated:

- **Inter-frame setup** (d_1) - Each physical processor core decodes $Q = 64$ LDPC frames in parallel to fully utilize the 512b SIMD units. All the values (channel, accumulators and messages) are stored on 8 bits.
- **Inter-frame with memory compression setup** - (d_2). Each physical processor core process $Q = 64$ frames in parallel. The channel values are compressed and stored using 1 bit whereas exchanged message values are compressed on 2 bits. The accumulator values are stored using 8 bits.
- **Intra-frame setup** (d_3) - A single LDPC frame is processed ($Q = 1$). All the values (channel, accumulators and messages) are stored on 8 bits.

Results presented in Table 2 show that in single core configuration, throughput from 77 Mbps up to 294 Mbps was measured for the d_1 implementation. The highest throughput was obtained when C_1 is decoded. Indeed, in this case, the inter-frame decoder has a small memory footprint (491 KB) that fills in L2/L3 caches. However, for long codes (C_2 and C_3), the throughput is divided up to 4× due to a memory footprint that grows up to 12125 KB. The d_2 implementation gives higher throughput for C_2 and C_3 codes due to memory compression that decreases the memory footprint to 4375 KB. However, the additional arithmetic and logical computations used to compress the information at runtime in d_2 makes it less efficient for C_1 code. At the opposite, the d_3 implementation provides the highest decoding throughput for C_2 and C_3 (from $\approx 1.2\times$ up to $\approx 2.4\times$ higher than d_2) with a maximum memory footprint of 194 KB. In parallel, the processing latency of d_3 implementation is about 99% shorter than d_2 ones. However, the usage rate of the SIMD units for d_3 is lower than 100% at runtime for C_1 LDPC code ($Z = 54$). Complex memory accesses and SIMD inefficiency in this case is not compensated by L1/L2 cache efficiency.

Experimentations were done when 20 or 40 processor cores are activated to check the scalability of the decoder implementations. The results show that the d_1 implementation is better in terms of throughput for C_1 code in all setups with speedup factors of $15\times$ and $25\times$. However, its performances fall down for C_2 and C_3 codes where the speedup factors are limited to $7\times$ to $10\times$. Indeed, even if the number of cores is increased from 20 to 40, a performance floor due to the maximal memory bandwidth appears. Memory bottleneck assertion is validated by the results obtained for d_2 implementation. Indeed, the d_2 implementation that executes memory compression over-classed d_1 one for the long frames (C_2 and C_3) codes. The throughput grows with the number of activated cores. However, this the performance improvement is not linear due to working frequency scaling. For d_2 decoder implementation, the measured speedups are $17\times$ and $24\times$ when 20 cores and 40 cores are activated, respectively. This phenomenon is due to the turbo-boost frequency scaling feature. In 20 and 40 core setups, the d_3 implementation offers the better performances and achieved up to 5446 Mbps. The speedup factors obtained in comparison with single core experiments are $14\times$ and $24\times$. At the same time, the working frequency reduction increases the processing latency by a factor of 2.

In parallel to the throughput and latency evaluation of the implementations, power and energy per bit comparisons were also done. Energy measurements are provided in Table 2. The reported energy consumption values include the CPU and the RAM consumptions. These values were obtained at runtime with the *turbostat* tool that captures the power consumption of sensors. The power consumption depends on the number of activated cores and the RAM usage rate. The power consumption results are equivalent to C_1 code for all implementations. However, for C_2 and C_3 codes, the d_1 and d_2 implementations have a higher power consumption ($\approx 40\%$) than d_3 due to their high usage rate of the RAM. Indeed, the RAM consumes up to 180 W. The energy per decoded-bit metric reinforces this performance gap about the inefficiency of the inter-frame parallelization scheme.

4.4 Comparison with FPGA-Based Decoder Implementations

Finally, the proposed Gallager-E LDPC decoder implementations on multi-core devices were compared with FPGA-based ones [16] to estimate the feature differences. Indeed, works presented in [16] details hardware optimized Gallager-E architectures implemented in a Zynq Ultrascale+ FPGA (xczu9eg-3ffvb1156e) based on architecture described in [12]. The number of hardware decoders allocated in the FPGA device varies according to the LDPC code concerned as reported in Table 3. This value was fixed to occupy the FPGA at 75% of its capacity to avoid place & route issues. The working frequencies post-PaR of the overall hardware experiments reach 500 MHz. Table 3 summarized for performance levels reached in terms of throughput, latency and energy.

The throughput and latency measurements first demonstrate that the FPGA solution provides 2× to 16× higher decoding throughput when a single processing core is considered on both platforms. This performance gap despite the favorable working frequency of the Xeon processor is due to the inefficiency of the x86 ISA. Indeed, a large set of basic computation requires on several clock cycles on the Xeon processor whereas in hardware they can be trivially implemented in one clock cycle. The observations related to the decoding latency between the two types of systems is in line with the observations made on throughput. The difference in terms of energy consumption per bit is more important. Factors are in the range 315× up to 830×. This is due to the high-power consumption of the Xeon processor when one core is active.

Table 3. Comparisons of software Gallager-E decoders with FPGA ones [16].

LDPC	Xeon implementation				FPGA implementation [16]			
code	# cores	Γ (Mbps)	Δ (μs)	E (nJ/bit)	#cores	Γ (Mbps)	Δ (μs)	E (nJ/bit)
C_1	1	220	6	819	1	620	2.1	1.94
C_2	1	180	113	956	1	3060	10.7	1.15
C_3	1	247	132	685	1	3020	6.8	1.13
C_1	40	5460	10	61	65	40300	2.1	0.9
C_2	40	4298	191	80	15	45900	10.7	0.97
C_3	40	5446	248	63	15	45300	6.8	0.89

Multi-core execution modifies the previous acknowledgment. The throughput difference is in this setup in range 5× to 11×. Indeed, the number of decoding cores that can be instantiated in the FPGA is lower than the number of cores in the Xeon processor, but at the same time, the Xeon working frequency is approximately halved. Consequently, the power consumption per decoded bit drops sharply for the Xeon solution because the power consumption is only double when 40 cores are activated compared to 1 core configuration and the throughput gain is thus over 20×. As a consequence, the difference in terms of power consumption between the FPGA implementation and the multi-core implementation varies finally from 52× to 82×.

This comparative section highlights the differences in terms of performance between dedicated architectures on FPGA targets and more flexible software solutions. As expected, dedicated hardware solutions are more energy efficient. However the gap with optimized software-based implementations that are faster to develop constantly decreases.

5 Conclusion

In this paper, three parallelized software LDPC decoder implementations are detailed. These software implementations implement the Gallager-E decoding algorithm which is efficient for hard input decoding of LDPC codes contrary to related work that manage soft-input values. The parallelization scheme applied and arithmetic optimizations used to implement this algorithm on INTEL Xeon multi-core target are detailed. Throughput up to 7,5 GBps is reported when 10 decoding iterations are executed. Finally, a comparison of throughput, latency and power measurements is done with FPGA-based implementation to highlight the efficiency of the proposer decoder implementations.

References

1. Andrade, J., Falcao, G., Silva, V., Sousa, L.: A survey on programmable LDPC decoders. IEEE Access **4**, 6704–6718 (2016)
2. Le Gal, B., Jego, C.: High-throughput multi-core LDPC decoders based on x86 processor. IEEE Trans. Parallel Distrib. Syst. **27**(5), 1373–1386 (2016)
3. Checko, A., et al.: Cloud RAN for mobile networks - a technology overview. IEEE Commun. Surv. Tutorials **17**(1), 405–426 (2015)
4. Chen, J., Fossorier, M.: Near optimum universal belief propagation based decoding of low-density parity check codes. IEEE Trans. Commun. **50**(3), 406–414 (2002)
5. Gallager, R.: Low density parity-check codes. IRE Trans. Inform. Theory **8**, 21–28 (1962)
6. Ghaffari, F., et al.: Efficient FPGA implementation of probabilistic Gallager B LDPC decoder. In: Proceedings of ICECS, pp. 178–181, December 2017
7. Ghaffari, F., Vasic, B.: Probabilistic gradient descent bit-flipping decoders for flash memory channels. In: Proceedings of ISCAS, pp. 1–5, May 2018
8. Giard, P., Sarkis, G., Leroux, C., Thibeault, C., Gross, W.J.: Low-latency software polar decoders. J. Sig. Process. Syst. **90**, 761–775 (2016)
9. Grayver, E.: Implementing Software Defined Radio. Springer, New York (2013). https://doi.org/10.1007/978-1-4419-9332-8
10. Le, K., Ghaffari, F., Kessal, L., Declercq, D., Boutillon, E., Winstead, C.: A probabilistic parallel bit-flipping decoder for low-density parity-check codes. IEEE Trans. Circuits Syst. I Regul. Pap. **66**(1), 403–416 (2018)
11. Le Gal, B., Jego, C.: Low-latency software LDPC decoders for x86 multi-core devices. In: Proceedings of SiPS (2017)
12. Le Gal, B., Jego, C., Leroux, C.: A flexible NISC-based LDPC decoder. IEEE Trans. Sig. Process. **62**(10), 2469–2479 (2014)
13. Marchand, C., Boutillon, E.: LDPC decoder architecture for DVB-S2 and DVB-S2x standards. In: Proceedings of SiPS, pp. 1–5, October 2015
14. Mitzenmacher, M.: A note on low density parity check codes for erasures and errors. SRC Technical Note 1998-017 (1998)
15. Pignoly, V., et al.: High data rate and flexible hardware QC-LDPC decoder for satellite optical communications. In: Proceedings of ISTC, pp. 1–5, December 2018
16. Pignoly, V., Le Gal, B., Jégo, C., Gadat, B.: Horizontal layered Gallager decoding of low-density parity-check codes for wireless up-link optical space communication. In: Proceedings of the ICECS, Glasgow, Scotland, 23–25 November 2020

17. Richardson, T.J., Urbanke, R.L.: The capacity of low-density parity-check codes under message-passing decoding. IEEE Trans. Inf. Theory **47**, 599–618 (2001)
18. Roberts, M.K., Anguraj, P.: A comparative review of recent advances in decoding algorithms for Low-Density Parity-Check (LDPC) codes and their applications. Arch. Comput. Methods Eng. **28**, 2225–2251 (2020)
19. Unal, B., Ghaffari, F., Akoglu, A., Declercq, D., Vasić, B.: Analysis and implementation of resource efficient probabilistic Gallager B LDPC decoder. In: Proceedings of NEWCAS, pp. 333–336, June 2017
20. Wadayama, T., Nakamura, K., Yagita, M., Funahashi, Y., Usami, S., Takumi, I.: Gradient descent bit flipping algorithms for decoding LDPC codes. IEEE Trans. Commun. **58**(6), 1610–1614 (2010)
21. Wubben, D., et al.: Benefits and impact of cloud computing on 5G signal processing: flexible centralization through cloud-RAN. IEEE Sig. Process. Mag. **31**(6), 35–44 (2014)

Low Latency Architecture Design for Decoding 5G NR Polar Codes

Oualid Mouhoubi(✉), Charbel Abdel Nour, and Amer Baghdadi

IMT Atlantique, Lab-STICC, UMR CNRS 6285, 29238 Brest, France
{oualid.mouhoubi,charbel.abdelnour,amer.baghdadi}@imt-atlantique.fr

Abstract. Polar codes have been adopted as the coding scheme in the control channel of the 3rd Generation Partnership Project (3GPP) New Radio (NR) standard for 5G. However, the challenging requirements introduced by the 5G control channel in terms of block length and code rate flexibility render unsuitable most of the published hardware polar decoder implementations. Indeed, these latter focused mainly on successive cancellation decoders with ultra-high throughput, limited flexibility and error correction capabilities. With stringent constraints on end-to-end delay and error correction, the 5G NR context steers towards low-latency list-based decoder architectures. In this context, we propose an original flexible list-based hardware architecture for decoding 5G NR polar codes that can support all the frame sizes and code rates defined in 3GPP with a worst-case decoding latency below 24 µs.

Keywords: 5G · polar codes · successive-cancellation decoding · list decoding · low latency · hardware design

1 Introduction

Proposed in the last few years, Polar codes have been shown to achieve channel capacity in binary discrete memoryless channels, when the codeword length N tends to infinity, using the low complexity successive-cancellation (SC) algorithm [3]. However, performance starts to degrade at practical code lengths. Hence, list-augmented SC decoding (SCL), where a list of L candidate codewords are considered during decoding [20] and the most reliable codeword is chosen at the end, improves the block error rate (BLER). Nonetheless, this comes at the cost of additional latency, chip area occupation and reduction in throughput for hardware implementations. In addition, the BLER of the SCL decoder can be further improved by appending a cyclic redundancy check (CRC) to select the most reliable codeword [15].

Being the cornerstone for all polar decoder types, several works have targeted the improvement of throughput and latency of the SC decoder through the proposal of solutions at both hardware design and algorithmic levels [4,16,19,22,23]. Through the introduction of specific decoders for constituent codes, simplified successive cancellation decoding (SSC) was proposed in [2] as an alternative

K. Desnos and S. Pertuz (Eds.): DASIP 2022, LNCS 13425, pp. 16–28, 2022.
https://doi.org/10.1007/978-3-031-12748-9_2

solution to reduce the latency. This latest work was then extended to the SCL decoder in [9,18].

Polar codes have been adopted for the control channel of the enhanced mobile broadband (eMBB) service of the 3GPP's 5G new radio (NR) standard. The control channel imposes block length and code rate flexibility levels far beyond previously published polar code designs. Moreover, low BLER and low hardware complexity added to a processing throughput and latency of 10 s of Mbps and 10 s of μs respectively are required [5].

Motivated by the need to provide a hardware-efficient polar decoder that supports the required flexibility and latency levels for 5G NR, we propose in this paper a novel hardware architecture targeting FPGA devices and offering the following features:

- Downlink and uplink 5G NR control channel compliance with full rate and frame size support ranging from $N = 32$ to $N = 1024$ bits.
- CRC-aided SCL decoder with a semi-parallel architecture [10] for best performance.
- Dedicated specific constituent-code decoders for reduced latency.
- Measured FPGA-based worst-case decoding latency of 23.91 μs, in compliance with target 5G NR constraints.
- Favourable comparison with previously-published designs.

The above-mentioned features were obtained thanks to the following hardware architecture-related contributions:

- Proposal and implementation of on-the-fly identifier for the number of constituent codes in addition to their type and length.
- Introduction of an original special node list decoder capable of decoding all identified constituent-code types.

The remainder of this paper is organized as follows. Section 2 provides a brief overview on polar codes and their decoding algorithms including fast decoding techniques used by our decoder. The proposed on-the-fly identifier and special node list decoder architecture are described in Sect. 3. Section 4 presents the implementation results and comparisons. Finally, Sect. 5 concludes the paper.

2 Preliminaries

2.1 Polar Codes

Polar codes apply the channel polarization transform that divides the bit-channels to either perfect or completely noisy channels. Then the polar code allocates information bits to the K most reliable bit-channels while the remaining bits are frozen, i.e., they are all set to a known value, usually '0'. For a codeword length $N = 2^n$, $n \geq 1$, a (N, K) polar code is a block code with K input bits and N output bits whose generator matrix G is the $n-$th Kronecker power of matrix $F = \begin{bmatrix} 1 & 0 \\ 1 & 1 \end{bmatrix}$, i.e., $G_N = F^{\otimes n}$. The encoding process is performed

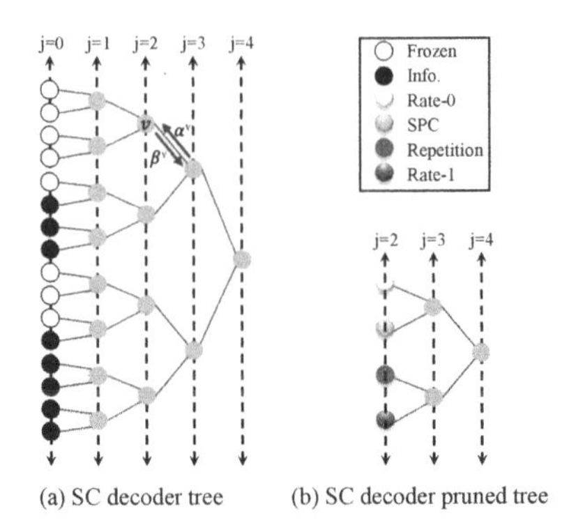

Fig. 1. SC based decoder tree and its corresponding pruned tree of (16, 8) polar code.

by the matrix multiplication $x = u.G$ where $u = (u_0, u_1, \dots, u_{N-1})$ stands for the sequence input vector consisting of information bits and frozen bits and $x = (x_0, x_1, \dots, x_{N-1})$ stands for the encoded vector.

With stringent constraints on rate flexibility and low decoding latency, polar codes were chosen in 5G NR to encode the uplink and the downlink control information over the physical uplink control/shared channels (PUCCH/PUSCH) and the physical downlink control/broadcast channels (PDCCH/PBCH). Therefore, they are required to support a wide range of information block length, encoded block and mother polar code lengths. CRC bits are appended to the information sequence. Three different CRC generator polynomials were carefully chosen for the purpose of improving the error correction performance. Depending on the physical control channels, CRC are differently initialized, scrambled and interleaved. They are allowed to trigger the end of the decoding process if the CRC check fails particularly during blind decoding. In addition to that, 3GPP decided to integrate in uplink two types of Parity Check (PC) bits in the middle of the encoded block. A so-called universal reliability sequence is used to determine the set of the frozen, information, CRC and PC bit positions for each polar code considered in 5G. With the intention of achieving the desired code rate R, reordering the N encoded bits or improving the error correction capability of polar codes, rate matching is the final step introduced in the encoding process used in the 5G NR control channels [1].

2.2 Successive-Cancellation Decoding Based Algorithms

The decoding of SC algorithms can be performed through a binary tree as illustrated in Fig. 1a for (16, 8) polar code. It consists of $\log_2 N+1$ stages where each stage j comprises $\frac{N}{2^j}$ nodes and each node represents a polar code of length 2^j. The top tree node at stage $j = \log_2 N$ includes the channel Log-Likelihood-Ratios (LLRs) and the final Partial Sums (PS). At leaf nodes, the frozen and information bits are represented by white and black circles respectively. A given node v receives α^v LLRs and produces β^v PS. Assuming that the processing of an activated stage can be performed in one clock cycle, the total number of time steps required to decode one frame is: $\mathbb{L}_{ref} = 2N - 2$. This corresponds to $\mathbb{L}_{ref} = 30$ in this example.

The major drawback of the SC algorithm resides in its inability to recover from wrong bit estimates, especially at the early stages of decoding. A SCL algorithm was proposed to avoid resorting to hard decisions when computing partial sums during the sequential decoding phase. Hard decisions are replaced by soft hypotheses for the error-prone bits identified by low reliability values. This leads to the simultaneous exploration of several codeword candidates or equivalently paths in the graph of Fig. 1a, each corresponding to one or more varying bit-hypotheses. Hence, for each bit u_i decoding step, both its possible values 0 and 1 are considered and $2L$ new candidate paths are explored. However, in order to break the exponential growth of the number of candidate paths, a subset L of the most likely paths is set to survive. The choice is made by selecting the L lowest path metric (PM) values. In terms of complexity, the SCL decoder can be seen as the concatenation of L competing SC decoders. Assuming that a path selection can be performed in one clock cycle, the latency of the SCL decoder becomes $T_{\text{SCL}}(N, K) = 2N - 2 + K$.

A simplified SC-based decoding algorithm (SSC) was presented in [2] in which the tree search is pruned. In fact, a tree with only frozen bits at leaves does not need to be traversed since its output is already known and is equal to an all-zero vector. This type of node is referred to as Rate-0. Moreover, a tree with only information bits can directly be decoded by applying a threshold decision at the root node. This type of node is referred to as Rate-1. Furthermore, the authors in [17] have identified two new types of nodes among the constituent codes of rate $0 \leq R \leq 1$. Hence, a repetition node (Rep) is a constituent code where all the bits are frozen except for the last one and the single parity check (SPC) node is a constituent code where at the exception of the first bit, all the bits are information. The pruned tree of the Fast-SSC decoder is shown in Fig. 1b where the four types of constituent codes are colored differently. This pruning technique was then extended to the SCL decoder. A hardware-friendly PM computation for Rate-0, Rep and Rate-1 nodes and for SPC node is presented in [8] and [7], respectively.

3 Proposed Decoder Architecture

3.1 Proposed Special Nodes List Decoding Module

Instead of decoding every constituent code type individually, as it is the case in [7], a special node list decoder which federates the common operations performed by the different special nodes is proposed. Considering that the tree should be traversed sequentially with a node by node processing schedule, special node decoders are able to share most of the memory and the computational resources. Hence, hardware duplication is avoided. The architecture of the special node list decoder is portrayed in Fig. 2.

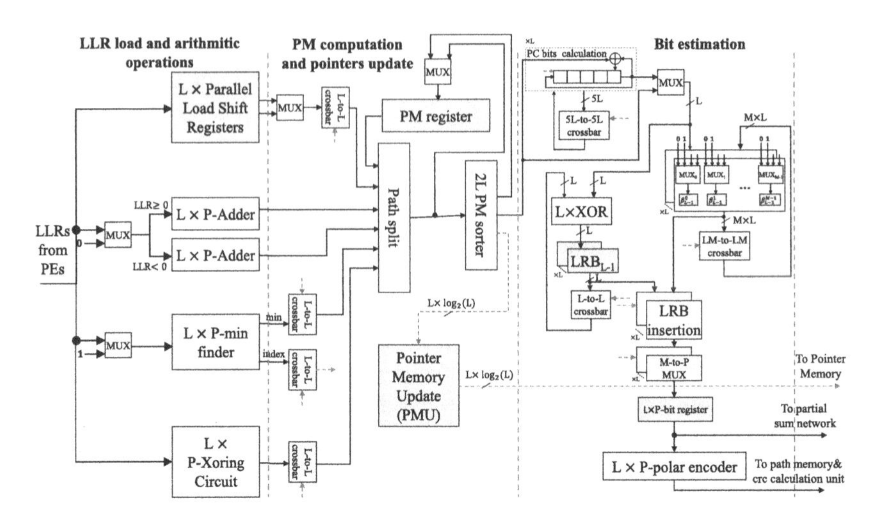

Fig. 2. Proposed special node decoder architecture supporting different special node types. (Color figure online)

Since the Rate-0 nodes comprise only frozen bits, no path splitting is needed to estimate the $N_{\text{Rate-0}}$ bits. Only summing up $N_{\text{Rate-0}}$ values is needed. A tree-structure fully parallel adder is preferred to this purpose in order to guarantee no extra decoding latency. The complexity and the critical path of this adder structure are high especially when the maximum size M defined for special nodes is large. As long as $P \leq M$, where P refers to the number of processing elements (PE) used by each path, the computations on the special node LLRs take multiple clock cycles and a maximum of P LLRs can be processed at a time. Therefore, smaller tree-structure adder can provide the same computational speed as the fully parallel adder. To this purpose, a sum-accumulate operation is used instead, by introducing an accumulator register at the output of the adder in order to limit the depth of the tree from $\log_2 M$ to $\log_2 P + 1$ stages and reduce the number of adders from $M - 1$ to P. As a result, $N_{\text{Rate-0}}$ values can be added up during $\lceil N_{\text{Rate-0}}/P \rceil$ clock cycles, where $\lceil x \rceil$ represents the closest integer value larger than x.

A Rep node of length N_{Rep} is identified by the presence of one information bit and $N_{\text{Rep}} - 1$ frozen bits. However a minimum of two steps are required for the estimate of the single information bit. The first one consists of updating the path metrics as part of the path fork process including both 0 and 1 decisions on the single information bit of the Rep node while the second consists of sorting the PMs of split paths to perform path selection. Unlike $N_{\text{Rate-0}}$ codes, two independent summations are needed by the Rep node. Indeed, according to whether the bit estimation is considered to be 0 or 1, Rep decoder has to add up all the negative valued or positive valued LLRs, respectively. For the purpose of performing both summations in parallel, the Rate-0 node adders are duplicated.

Rate-1 nodes comprise only information bits which are decoded one-by-one as in the SCL algorithm. For each bit estimate, paths are duplicated, sorted and some of them are discarded. Since more than one bit is decoded in a row without going back to PEs, the special node unit comprises a *Pointer Memory Update* (PMU) unit that keeps track of the surviving paths from the beginning of each Rate-1 and SPC nodes decoding.

A SPC node is identified by the presence of one frozen bit while all the remaining bits are information. The least-reliable bit (LRB) is found as a first step for decoding an SPC node. It corresponds to the minimum LLR value at the top of the SPC tree node. An accumulation-minimum-finder tree-based structure similar to the one used for the adder units of Rate-0 and Rep nodes with P comparators is used in this regard. The even-parity check evaluation, in turn, is performed through an accumulator-XOR tree-based structure using P XOR gates, while the evolving even-parity constraint is performed for each of the existing L paths after each bit estimate through L XOR gates and stored in dedicated LRB registers. The final value of the latter is retained to preserve the even-parity constraint and inserted within the estimated bit vectors directly in its appropriate location provided by the minimum value index found in the first step (green arrow in Fig. 2).

Except for a classical bit decoding case, more than one LLR are admitted by the special node decoders. To avoid complexity due to write/read operation, they are stored in a bank of parallel-load shift registers, as illustrated in Fig. 3. Assuming $P \leq M$, they are seen as an array of M Q_i-bits registers grouped in M/P independent columns which can be accessed independently thanks to M/P enable signals. Q_i is the quantization level of the internal LLRs. LLR's nodes are stored in the same order they are produced, starting from the rightmost column to the leftmost one during $\lceil \frac{2^j}{P} \rceil$ clock cycles. During the special node decoding phase, these registers are shifted both horizontally and vertically during the decoding phase of Rate-1 and SPC nodes as follows:

- Vertical shift: LLRs are accessed one-by-one by shifting the rightmost P registers from top to bottom.
- Horizontal shift: After $P-1$ vertical shift, the following P LLRs to access, when they exist, are obtained by shifting registers of the different columns horizontally from left to right.

Excluding the last loaded registers column, the remaining ones can accept their LLRs from either their adjacent one or from input. This is made possible thanks to $P \cdot \left(\frac{M}{P} - 1\right)$ 2-to-1 MUX. Since the least reliable bit in an SPC node is decoded first, its LLR value should be skipped whenever it is encountered. To do this, the LLR bank register in Fig. 3 is designed to output two adjacent LLR values instead of one. A MUX is used to select the output based on the least reliable bit index. When the PC bits are used in uplink, they are decoded using the length-5 cyclic shift register (Fig. 2). To support path competition, the generated intermediate paths read their values, i.e., LLRs, even-parity check and minimums by means of crossbars. Newly estimated bits are temporarily stored in dedicated registers.

A straightforward copy operation allows, by means of crossbars, to move all the contents of each of the L surviving paths from a register to another after each path selection. The process of PM sorting and bit estimation is repeated $N_{\text{Rate-1}}$ or $N_{\text{SPC}} - 1$ times. The PM register is updated directly from the $2L$ sorter in the same clock cycle. When all bits are estimated at their top tree nodes, the source word bits are obtained through a polar encoder, the PMU updates the pointer memory and the search procedure of the next nodes in the polar code tree resumes. The partial sums needed to update LLRs are computed and the CRC check process is resumed. In order to avoid extra latency, PS and LLR computations are performed in parallel.

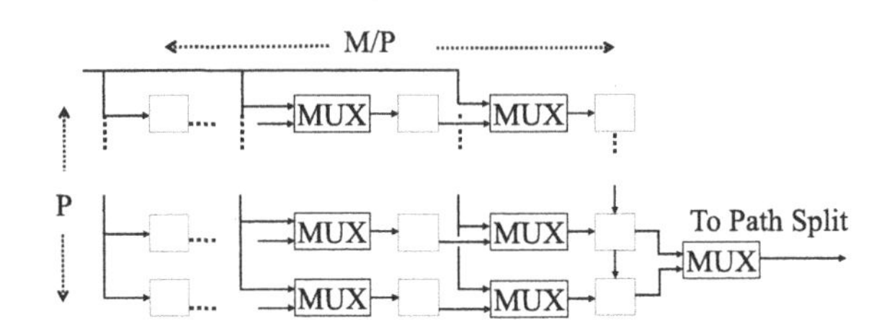

Fig. 3. Parallel-load shift registers of the top-node LLRs.

3.2 Memory Structure

LLR Memory: The SC decoding relies on dedicated memories to keep the LLR values available for the computation units as long as they are needed. Therefore, channel input LLRs are stored in a $N \times Q_c$ bits register, where Q_c is the quantization level of the channel LLRs. However, internal LLRs are stored in a dual-port RAM-based memory configured with one write port and one read port. The total number of LLR updates at decoding stage j is equal to 2^j. However, only P LLR updates are allowed to be performed at once. Thereby, for stages j where $2^j > 2P$, a total of $2^j/(2P)$ time steps are needed before proceeding to the lower stage while only one time step is needed for the other stages. Given that the computed LLRs in PEs have to follow the bit-reversed indexing scheme, the produced LLRs need a certain reordering before being mapped back to PEs. To avoid the need for multiplexing them, the equivalent operation is directly embodied in the control unit and two RAMs instead of one, each of width PQ_i, are implemented for each path l. For higher stages, LLRs are first stored in one RAM during half of the $2^j/(2P)$ time period required for computation before storing the remaining produced LLRs during the second half of the time period in the other one. However, for lower stages, where LLRs are produced during one time period, LLRs contiguity is ensured by adding a specific permutation network. In addition, a $L \times PQ_i$-bit buffer is used in order to process these generated LLRs directly during the following clock cycle.

Pointer Memory: The SCL decoder can also be seen as L SC decoder cores working in parallel with their own LLRs and memory resources. Nonetheless, the cores may share the same LLRs at some stages due to path competition. Thereby, to avoid copying the shared LLRs, a memory pointer is introduced and stored instead. Hence, each SC decoder core can read its inputs from one of the L RAM memories thanks to a permutation network. The used memory pointer is a $(\log_2 N - 1) \times L$ register array, where register at row j and column l stores a pointer of $\log_2 L$ bits indicating the index of the RAM where LLRs of path l at decoding stage j are stored. The pointers are updated in two different situations. First during path duplication/discarding. Second, to reset them to their initial values when the stages at which they are processed is activated.

Path Memory: A LN array register is dedicated to store the values of the L codeword bits after the PMs have been sorted and the surviving paths identified. Being mutually independent, all registers can be accessed simultaneously. This is made possible thanks to a N-bit enable vector. When a path l needs to be duplicated, all its so far decoded bits are copied to the registers that have been freed, due to path discarding, by means of N L-to-1 MUX. Enable signals are generated by a $\log_2 N$-to-N decoder which takes the currently decoding information bit index as input. N MUXes are used to select between the copy/store operation to perform and P-to-P crossbars are used to manage the flexibility in decoding variable special node lengths.

3.3 Proposed On-the-Fly Rate-Flexible Decoding of Polar Codes

To apply dedicated special node decoders for the 5G NR polar code, the types and sizes of the nodes within the current frame need to be provided. The large flexibility in frame and code rate ranges leads to a large number of different frozen bit sets. This increases the number of different special nodes, making it difficult to store in memory their number and positions. To tackle this issue, we propose a new method to identify the different special node structures (Rate-0, Rep, SPC and Rate-1) on-the-fly directly in hardware without the need to store any list in memory. Their determination is done from the frozen set and by merging different lower size special nodes into a new special node type of larger size. For hardware optimization purposes, we define a new special node type corresponding to the only case where two bits from the frozen set do not represent any of the particular constituent code listed above, i.e., the sequence {*information, frozen*} and is called *No type* node. Henceforth, a unique binary sequence refers to both *No type* and SPC nodes. However, assuming having additional information on their size, these two special nodes can not be overlapped since SPC nodes start to be considered from length-4 patterns.

Starting from a given frozen set, the structure of Fig. 4a is capable of determining the different special node types of length-2. For higher length nodes, the structure of Fig. 4b intends to merge different lower size special nodes into a new special node type of larger size based on a sequence vector indicating the type of

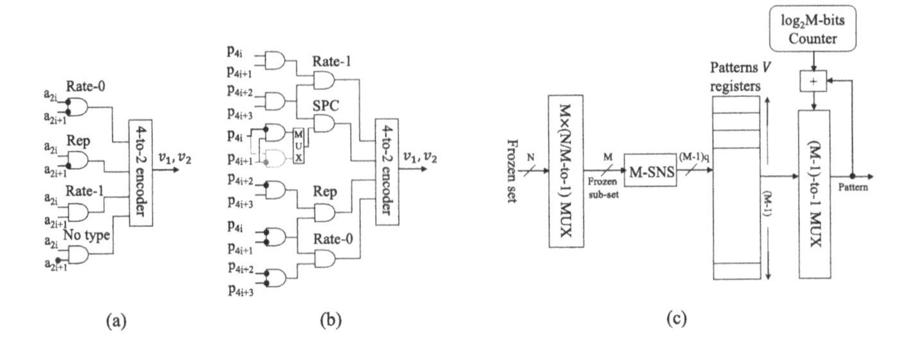

Fig. 4. Proposed architecture for identification of constituent codes of different lengths: (a) Length two elementary module identifier, (b) Length four and above elementary module identifier, (c) Architecture of the special node identifier.

the nodes to merge. With one difference, a SPC node of length four is obtained by merging a Rep and Rate-1 nodes of length two (highlighted in red) while an SPC node of length greater than four is obtained by merging another SPC and Rate-1 nodes, hence the presence of the MUX. Therefore, the two structures of Fig. 4a and Fig. 4b represent the building blocks of the identifier intended to determine the different supported special node types of any length.

The special and non-special nodes identified are stored together in register arrays in the same order as they are searched during the decoding process. This reduces the complexity of multiplexing the pattern compared to the case where only valid ones are retained. Since the SC decoder is sequential, the special node vectors V (Fig. 4c) are read from the register one at a time through a $\frac{N}{M} \cdot (M-1)$ to-1 MUX. A $\log_2 N$-bit counter is used to generate the register-based memory addresses. At each special node iteration search, the next pattern to read from the registers array depends on the size and address of the previous one. The complexity of the proposed architecture increases linearly with code length N. But since the special nodes are searched only upon request, a low complexity serial implementation is favoured. To do this, a N-to-M multiplexer is used to process serially the frozen set bits in groups of M. The architecture of a serial search of special nodes is depicted in Fig. 4c.

4 Synthesis Results and Comparisons

As a proof of concept, a decoder architecture fully compliant with the 5G NR polar code has been designed. The proposed architecture is generic with respect to the parameters P, L and M. An analysis of the impact of main code and decoder design parameters, including P and M, on the latency and the hardware complexity is proposed in [14]. In addition to the proposed special node list decoding and special node identification unit, the architecture includes: $L \times P$ *Processing Elements (PE)* to compute and update the LLRs, *Multi-bit Partial*

Table 1. Average and maximum latency measured by the proposed decoder.

			Worst-case latency		Average Latency	
Physical control channel	N	# code	cc	[μs]	cc	[μs]
Downlink	64	435	146	1.35	121	1.11
	128	3451	284	2.63	207	1.91
	256	9452	483	4.47	376	3.47
	512	2286	780	**7.22**	719	6.65
Uplink	1024	3491	2583	**23.91**	1914	17.72

Sum Network to produce PS, *CRC bits calculation unit* to perform CRC check operations in the case of multi-bit decoding and which operates in parallel to the decoding process. The *Control unit* integrates a finite state machine designed to generate all the control signals needed for the different decoder components. In order to manage candidates competition, crossbars were also designed whether for copying data to some freed memory locations or to ensure a correct routing of data as a result of the use of pointer memories.

The devised architecture has been described in VHDL. The FPGA used for synthesis was a Xilinx Virtex 7-xc7vx485t device. The number of bits used to represent internal LLR, PM and channel LLR values is $Q_i = 6$, $Q_p = 7$ and $Q_c = 4$, respectively. Eight PEs are instantiated in the architecture for each of the L paths, where $L = 8$. The maximum size of special nodes M is set to 32. Internal LLRs and partial sum bits are stored in the FPGA dual-port Block RAMs (BRAMs) while input LLRs and partial sum memory addresses are stored using look-up tables (LUTs). Decoding latency is the main critical performance metric when considering the 5G NR polar codes that protect the control channel. A target end-to-end latency of 0.5 ms implies a physical layer latency of 50 µs [6,13]. Flexibility and low latency are the driving priority for the design of this channel decoder, which is the main demanding component of the physical layer.

The latency for decoding one codeword is not constant and is highly affected by the choice of the operating code length and code rate. In order to give a full analysis of the latency, all combinations of code lengths and code rates ranging from $R = 1/8$ to $R = 5/6$ are evaluated through simulations. Table 1 provides the average and the worst-case latency for decoding one codeword. The latency is measured in number of clock cycles (cc) and in µs considering the maximum clock frequency $f_{\max} = 108$ MHz. The results show that the maximum latency recorded by the decoder is 23.91 µs which is 2.1 times lower than the physical layer latency constraint. This worst-case latency appears in the uplink scenario, whereas it is only 7.22 µs in the downlink.

Logic synthesis and performance results are summarized in Table 2. Implementations that are fully compatible with 5G code specifications, in terms of code structure and flexibility, allow for direct and fair comparisons. Therefore, the recently available Xilinx polar decoder [21] is the most relevant reference with respect to the available decoder designs in the literature. Performance results of Xilinx decoder are available, yet with no published details on the architecture. Compared to this decoder, our proposed architecture has 60% and 70%

Table 2. Comparison with several FPGA-based SCL Architectures.

Decoder	[12]	[11]	**This work**	Xilinx [21]		**This work**	
List size		4			8		
Algorithm	SCL	SSCL[1]	SSCL[2]	N/A		SSCL[2]	
Flexibility	Limited	Limited	High	High		High	
Quantization	(10,3)	NA	(6,7,4)	8		(6,7,4)	
# PE	64	64	8	NA		8	
$f_{\max}$ (MHz)	N/A	445.2	108	223		108	
FPGA Device	Stratix V	Stratix V	xc7vx485t	xc7vx485t		xc7vx485t	
ALMs/LUTs	101160	8146	**26049**	45569		**51262**	
FFs	13544	2862	**12603**	33063		**20270**	
BRAMs	0	0	**34**	51.5		**68**	
(N,K)	(1024,512)	(1024,512)	(1024,512)	(512,40)	(1024,512)	(512,40)	(1024,512)
Latency (μs)	4064[3]	1177	**10.73**	11.35	46.41	**4.76**	**18.37**
TP (Mbps)	N/A	0.452	47.7	28.68	88.4	8.4	34.4

[1] Stochastic SCL with 2-level decoding.
[2] Simplified SCL with four constituent codes.
[3] Number of clock cycles.

less decoding latency for the uplink (1024,512) and the downlink (512,40) polar codes, respectively. Our design consumes 38% less Flip-Flops (FFs), but 11% more LUTs and 25% more BRAMs. The throughput of the Xilinx polar decoder, however, does not compare favorably with our architecture, yet still compliant with the 5G NR requirement for this code used for the control channel.

In order to extend the comparison, we have considered recent designs that targeted FPGA implementation with similar code length, yet not compliant with 5G NR polar codes. For that, a second configuration of our proposed architecture has been designed and synthesized with $L = 4$ while keeping the number of PEs per list unchanged. Compared to the folding polar decoder of [12] that achieves comparable BLER, our decoder uses 74% less LUTs and 6% less FFs while decoding the (1024,512) code in less than half the time (converted in clock cycles). However, without any reported clock frequency, the latency comparison is not complete. Compared to the stochastic SCL decoder of [11] targeting wearable and IoT devices with strict hardware constraints, our decoder, supporting rate and frame size flexibility and providing better BLER performance, exhibits 109 times less latency while requiring 3.2 times and 4.4 times the numbers of LUTs and FFs, respectively.

5 Conclusion

In this paper we proposed an original hardware architecture for decoding the 5G NR polar codes of the uplink and the downlink control information channel. Thanks to a special node identifier, the proposed decoder continues to benefit from tree pruning techniques to speed-up the decoding whilst maintaining compliance with the 5G NR and the various defined combinations of code rate and code length. Hence, measured throughput and latency values of the proposed decoder obtained with FPGA target are able to meet 5G requirements. Moreover, synthesis results have shown a hardware efficiency that compares favourably with state-of-the-art FPGA implementations of polar decoders. Finally, the designed architecture reduces the decoding latency compared to the recently available Xilinx 5G polar decoder.

References

1. 3GPP: TS NR multiplexing and channel coding. Release 15 V15.6.0 TS, 38.212 (2019). https://portal.3gpp.org/desktopmodules/Specifications/SpecificationDetails.aspx?specificationId=3214
2. Alamdar-Yazdi, A., Kschischang, F.R.: A simplified successive-cancellation decoder for polar codes. IEEE Commun. Lett. **15**(12), 1378–1380 (2011)
3. Arikan, E.: Channel polarization: a method for constructing capacity-achieving codes for symmetric binary-input memoryless channels. IEEE Trans. Inf. Theory **55**(7), 3051–3073 (2009)
4. Bian, X., Dai, J., Niu, K., He, Z.: A low-latency SC polar decoder based on the sequential logic optimization. In: International Symposium on Wireless Communication Systems (ISWCS) (2018)
5. Egilmez, Z.B.K., Xiang, L., Maunder, R.G., Hanzo, L.: The development, operation and performance of the 5G polar codes. IEEE Commun. Surv. Tutorials **22**(1), 96–122 (2019)
6. Fettweis, G.P.: The tactile internet: applications and challenges. IEEE Veh. Technol. Mag. **9**(1), 64–70 (2014)
7. Hashemi, S.A., Condo, C., Gross, W.J.: A fast polar code list decoder architecture based on sphere decoding. IEEE Trans. Circuits Syst. I Regul. Pap. **63**(12), 2368–2380 (2016)
8. Hashemi, S.A., Condo, C., Gross, W.J.: Simplified successive-cancellation list decoding of polar codes. In: IEEE International Symposium on Information Theory (ISIT), pp. 815–819 (2016). https://doi.org/10.1109/ISIT.2016.7541412
9. Hashemi, S.A., Condo, C., Gross, W.J.: Fast simplified successive-cancellation list decoding of polar codes. In: IEEE Wireless Communications and Networking Conference Workshops (WCNCW), pp. 1–6. IEEE (2017)
10. Leroux, C., Raymond, A.J., Sarkis, G., Gross, W.J.: A semi-parallel successive-cancellation decoder for polar codes. IEEE Trans. Sig. Process. **61**(2), 289–299 (2013). https://doi.org/10.1109/TSP.2012.2223693
11. Liang, X., Wang, H., Shen, Y., Zhang, Z., You, X., Zhang, C.: Efficient stochastic successive cancellation list decoder for polar codes. Sci. China Inf. Sci. **63**(10), 1–19 (2020). https://doi.org/10.1007/s11432-019-2924-6
12. Liang, X., Yang, J., Zhang, C., Song, W., You, X.: Hardware efficient and low-latency CA-SCL decoder based on distributed sorting. In: 2016 IEEE Global Communications Conference (GLOBECOM), pp. 1–6 (2016)
13. Maunder, R.G.: The 5G channel code contenders. ACCELERCOMM White Paper, pp. 1–13 (2016)
14. Mouhoubi, O., Abdel Nour, C., Baghdadi, A.: On the latency and complexity of semi-parallel decoding architectures for 5G NR polar codes. In: 11th International Symposium on Signal, Image, Video and Communications (ISIVC) (2022)

15. Niu, K., Chen, K.: CRC-aided decoding of polar codes. IEEE Commun. Lett. **16**(10), 1668–1671 (2012)
16. Roy, S.J., Lakshminarayanan, G., Ko, S.B.: High speed architecture for successive cancellation decoder with split-g node block. IEEE Embed. Sys. Lett. **13**, 118–121 (2020)
17. Sarkis, G., Giard, P., Vardy, A., Thibeault, C., Gross, W.J.: Fast polar decoders: algorithm and implementation. IEEE J. Sel. Areas Commun. **32**(5), 946–957 (2014). https://doi.org/10.1109/JSAC.2014.140514
18. Sarkis, G., Giard, P., Vardy, A., Thibeault, C., Gross, W.J.: Fast list decoders for polar codes. IEEE J. Sel. Areas Commun. **34**(2), 318–328 (2016)
19. Shrestha, R., Sahoo, A.: High-speed and hardware-efficient successive cancellation polar-decoder. IEEE Trans. Circuits Syst. II **66**(7), 1144–1148 (2018)
20. Tal, I., Vardy, A.: List decoding of polar codes. IEEE Trans. Inf. Theory **61**(5), 2213–2226 (2015)
21. Xilinx: IP Polar Encoder/Decoder (2021). https://www.xilinx.com/products/intellectual-property/ef-di-polar-enc-dec.html#overview
22. Yuan, B., Parhi, K.K.: Low-latency successive-cancellation polar decoder architectures using 2-bit decoding. IEEE Trans. Circuits Syst. I Regul. Pap. **61**(4), 1241–1254 (2013)
23. Zhang, C., Parhi, K.K.: Low-latency sequential and overlapped architectures for successive cancellation polar decoder. IEEE Trans. Sig. Process. **61**(10), 2429–2441 (2013)

Efficient Software and Hardware Implementations of a QCSP Communication System

Camille Monière[1,2(✉)], Bertrand Le Gal[2], and Emmanuel Boutillon[1]

[1] Lab-STICC, Université de Bretagne Sud, 56100 Lorient, France
{camille.moniere,emmanuel.boutillon}@univ-ubs.fr
[2] IMS, Bordeaux-INP, 33400 Talence, France
bertrand.legal@ims-bordeaux.fr

Abstract. In wireless communications, frame detection and synchronization are usually performed using a preamble, consuming bandwidth and resources that are not negligible for small packets. Recently, a new kind of preamble-free frame called Quasi Cyclic Small Packet (QCSP) have been proposed. This paper studies the implementation of QCSP transmission, both at the transmitter side and the receiver side. For the latter, only detection, the most consuming task, is considered. Different parallelism levels and implementation strategies are detailed for both software and hardware implementations. Several trade-offs between throughput and resource usage are also discussed. Finally, the paper demonstrates that the emission/reception process of a QCSP frame is feasible at low hardware cost, which make the QCSP frame very attractive for Low Power Wide Area Networks (LPWAN).

Keywords: Real-Time Implementation · CCSK · Small Packets · Hardware · Software · Low Power Wide Area Network

1 Introduction

Transmitting small amount of data in an unsupervised communication network is a real challenge, especially for Internet of Things (IoT) devices. For such devices, a frame usually consists on a payload with additional redundancy (for error-tolerance), preceded by a standardized preamble which help to the synchronization task. Unfortunately, for short packets, the preamble size is no longer negligible compared to the payload size. Indeed, if preamble-based methods allow to greatly simplify the receiver complexity, thanks to the known parts of the received waveform, the energy used during its transmission is simply wasted from the communication point of view [13]. As an example, in IoT context such as Massive Ultra Reliable Low Latency context, preambles already consume a significant amount of bandwidth and energy [8]. Preamble-less strategies for the transmission of short packets exist in the literature [3,4,16]. However, their efficiency has been demonstrated only for positive SNR values.

K. Desnos and S. Pertuz (Eds.): DASIP 2022, LNCS 13425, pp. 29–41, 2022.
https://doi.org/10.1007/978-3-031-12748-9_3

A recently introduced preamble-less frame called Quasi Cyclic Small Packet (QCSP) [5,15] have been proven to work at very low SNR (until −10 dB) solving this issue. A QCSP frame is based on the association of a Cyclic Code Shift Keying (CCSK) modulation and a Non-Binary Error Control Code (NB-ECC). It is shown that a Miss Detection Probability ($\mathcal{P}_{MD}$) of 10^{-4} for a False Alarm Probability ($\mathcal{P}_{FA}$) below 10^{-6} can be reached. Moreover, the proposed algorithm is robust to frequency inaccuracies and receiver gain variations.

From an implementation point of view, the QCSP waveform has a low computational complexity and a small memory footprint at transmitter side. This feature is needed by IoT devices, that are constrained by power usage. The drawback is the increase in complexity of the detection and synchronization tasks in reception. Although expected, given the lack of a preamble, it makes the implementation of a real-time receiver challenging.

In [5], the authors described the principle of sliding windows for QCSP frame detection and compared the new method performance with the classical FFT method. Early results of software implementation were also given to show the relevance of the method. This article extends this work, focusing on the in-depth study of existing implementations of QCSP transmitter and detector, rather than algorithmic description. The article is structured as follows. Section 2 first presents the system model, and then briefly exposes the detection methods presented in [5]. Section 3 details current architecture variations, estimating and comparing their respective complexity. In Sect. 4, the analysis of the parallelization schemes applicable to each evaluated detection approach is reported. The results coming from transmitter and detector implementation on CPUs and FPGA are reported and explained in Sect. 5, with the various tweaks and optimizations involved. Finally, conclusions and perspectives are reported in Sect. 6.

2 System Model

This section reviews the communication system and the associated detection method presented in [5]. First, the QCSP frame is introduced. Then, transmitter and receiver sides of the communication system are described.

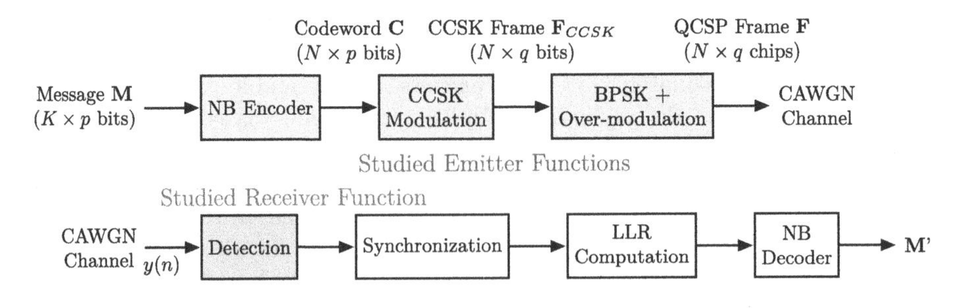

Fig. 1. QCSP communication system model (Color figure online)

The communication system is depicted in Fig. 1. The message sent $\mathbf{M}$ (K symbols of p bits, each symbol being in $[0, 1, \ldots, q = 2^p]$) by the transmitter is first coded using an Non-Binary Low Density Parity Check (NB-LDPC) encoder into a codeword $\mathbf{C}$ (N symbols of p bits). Then, the NB symbols are CCSK modulated using a binary pseudo random noise sequence $\mathbf{P}_0$ circularly shifted (k-shifted $\mathbf{P}_0$ being denoted $\mathbf{P}_k$) [7]. The resulting CCSK frame is modulated by Binary Phase-Shift Keying (BPSK) and over-modulated [15] allowing blind symbol synchronization. This creates a QCSP frame $\mathbf{F}$, which filtered by a half raised cosine filter, resulting in an up-sampling by a factor $\mathcal{O}$, and sent to an RF device for transmission. During reception, samples are filtered by the same filter, then $\mathcal{O}$ detectors works in parallel to allow future decimation. Each one use CCSK based detection, described in [14] and summarized in the next subsection. If a detection occurs, data associated with the highest-scoring over-sampling hypothesis are given to synchronization, which removes any frequency, time or phase inaccuracies. Log-Likelihood Ratios (LLRs) are computed from the resulted synchronized frame during CCSK demodulation, which is the correlation with $\mathbf{P}_0$ [2] (operator $\star$). Finally, they are fed to an NB-LDPC decoder to retrieve the received message.

This paper will focus on the implementation of a full real-time transmitter (in green in Fig. 1) and of real-time detectors (in red in Fig. 1).

CCSK-based detection mainly consists in comparing a score function to a threshold value U_0. The score is computed using the last $N \times q$ received samples (i.e. the length of a frame) at time n, equally divided in N sub-vectors $\mathbf{Y}_n$ (vector of $y(n-i)$ for $i \in [0, 1, \ldots, q-1]$) of q samples. If the score value exceed the threshold U_0, a new frame arrival is assumed.

The score function S_n^ω corresponds to a filter output that is maximized for a frame that arrived at time n with a frequency offset $f = \frac{\omega}{2\pi q}$. The first step is to mitigate the frequency offset by multiplying term by term (operator $\odot$) $\mathbf{Y}_n$ with rotation vector $\mathbf{\Gamma}^\omega = \{1, e^{-j\frac{\omega}{q}}, e^{-j\frac{2\omega}{q}}, \ldots, e^{-j\frac{(q-1)\omega}{q}}\}$ (i.e. a pure complex sinusoidal of frequency $-f$). This operation gives $\mathbf{Y}_n^\omega = \mathbf{Y}_n \odot \mathbf{\Gamma}^\omega$. A correlation vector $\mathbf{L}_n^\omega$ is then computed as

$$\mathbf{L}_n^\omega = (\mathbf{Y}_n \odot \mathbf{\Gamma}^\omega) \star \mathbf{P}_0. \tag{1}$$

This vector is similar to an attempt of CCSK demodulation. The maximum normalized absolute value of $\mathbf{L}_n^\omega$ (denoted M_n^ω) is taken as an indicator of the demodulation success. S_n^ω results from the accumulation of M_n^ω through an averaging filter the size of a frame.

Since the resulting score is compared to a threshold U_0, and provided that the threshold U is adapted, the square roots involved in the calculus of in [5] can be removed, thus simplifying the score function. Therefore, M_n^ω results from

$$M_n^\omega = \frac{\max\{|L_n^\omega(i)|^2, i = 0, 1, \ldots q-1\}}{\sum_{i=0}^{q-1} |y(n-i)|^2}, \tag{2}$$

and the score function is

$$S_n^\omega = S_{n-q}^\omega + M_n^\omega - M_{n-Nq}^\omega. \tag{3}$$

S_n^ω being the score associated to the last $N \times q$ samples, it must be computed at least every q received samples, but can be computed up to q times. It leads to the introduction of the parameter p_Δ, power of 2 indicating the number of score values computed every q samples for a rotation ω (i.e. $p_\Delta = 1, 2, 4, \ldots, q$). It should be noted that it also reduces the memory usage, since only $N \times p_\Delta$ values of M_n^ω are needed. Finally, to ensure frequency-error tolerance, multiple score S_n^ω for different value of ω are computed in parallel. The number of rotation hypotheses tested in parallel from $-\pi$ to π is denoted p_ω, ranging from 1 to theoretically any natural integer, but limited to 8 in practice, the detection performance gain being minimal in regard to the resulting cost in complexity for higher values [5]. The overall architecture that allows to compute a score S_n^ω is given in Fig. 2, and is called Score Processing Unit (SPU). The two existing SPU architectures are summarized in the next section.

3 Score Processing Units

The legacy method [14] to compute the correlation ($\star \mathbf{P}_0$ in Fig. 2) is depicted in Fig. 3. This method, although proven and well documented, is not well suited for continuous processing of incoming samples due to its computational complexity. Indeed, to achieve the best performances (i.e. $p_\Delta = q$), for each arriving sample, the q long FFT and IFFT must be reprocessed (except for the FFT of $\mathbf{P}_0$, which is constant and thus can be stored in memory). Lowering p_Δ may lessen the computation complexity, without impacting detection performances too much. In this case, a new correlation is produced every $\frac{q}{p_\Delta}$ new received data.

An optimized method to compute the correlations is introduced in [5]. This approach computes the correlation in the time domain, using

$$L_n^\omega(k) = L_{n-1}^\omega(k-1) + p_k d_n^\omega, \tag{4}$$

with $d_n^\omega = (y(n) - y(n-q)e^{j\frac{\omega}{q}})e^{-jn\frac{\omega}{q}}$ and $p_k = P_k(q-1)$, for $k = 0, 1, \ldots, q-1$. This new method, depicted in Fig. 4, is purely iterative, and uses the former correlation result to compute the new one. To this end, an accumulation window sliding through time is created. The q values of each correlation resulting from those of the previous one, correlations must be fully computed for every sample.

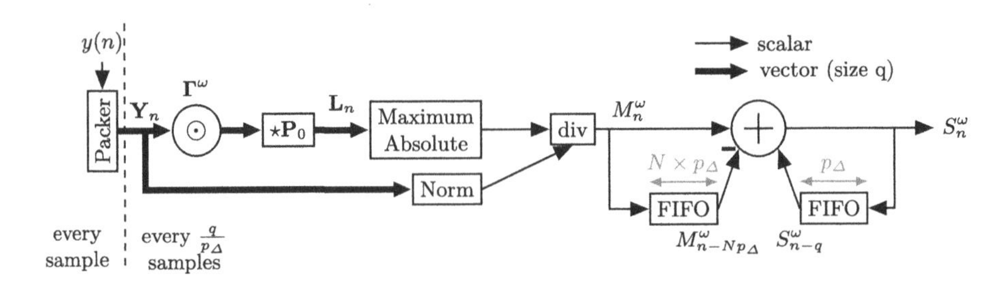

Fig. 2. Complete elementary score processing unit for a given rotation ω.

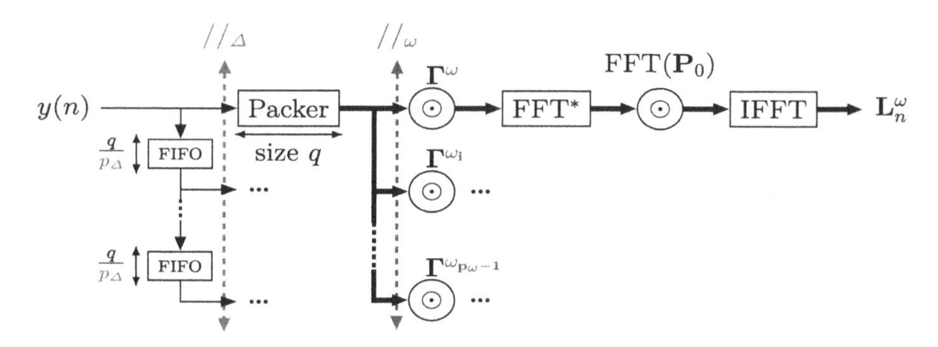

Fig. 3. FFT-based SPUs and their associated Δ-parallelism and ω-parallelism

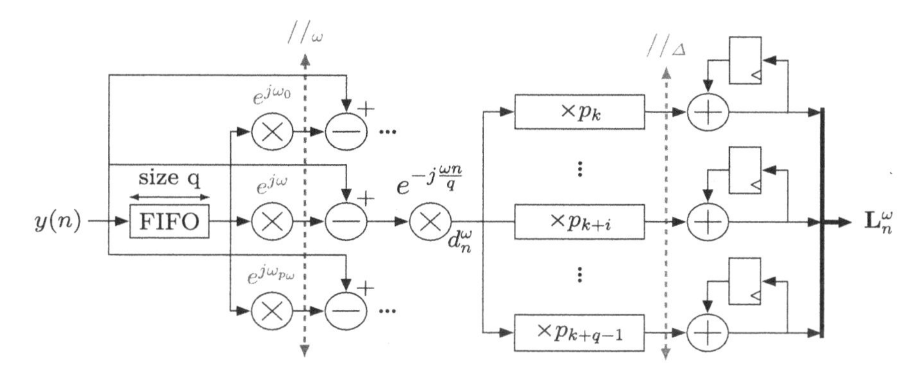

Fig. 4. Time sliding SPUs and their associated Δ-parallelism and ω-parallelism

Correlations being computed at each new sample, it results in a fixed p_Δ of q for time sliding SPUs.

The complexity of each correlation method is summarized in Table 1. These complexities are currently reported in terms of floating-point arithmetic/storage. All proposed method complexities are directly proportional to p_ω, so this parameter has not been included in Table 1.

Different setups were considered for the correlation methods. Indeed, FFT SPUs can either distribute or share the memory needed by packers. Since the first SPU already stores the q last samples, the second just need $\frac{q}{2}$ samples more, the third and the fourth need $\frac{q}{4}$ in addition, and so on. Concerning the time sliding method, the last multiplication can be simplified by a simple conditional negate operation, $\mathbf{P}_0$ being a $\{-1+1\}$ binary sequence in the temporal domain. It reduces by a factor q the number of complex multiplications (this method reported as Time Sliding* in Table 1).

According to the values reported in Table 1, the time sliding architecture has the lowest complexities when $p_\Delta = q$ is considered (the best detection case). However, p_Δ can be lowered when using FFTs, which can result in a lower complexity at the cost of low impact on detection performances. Exploration of impact the of p_Δ value is detailed in [5]. At this point, no solution seems clearly better than the other for real time implementation, because even if less computations are required by the time sliding method, it also offers a lower-level parallelism. This aspect is discussed in the following section.

4 Parallelization Strategies Applied to Detection Task

The detection task is the most demanding task of the receiver. Indeed, it processes all received samples with complex detection algorithms. To achieve real-time performances, the two detection methods must be parallelized to benefit from multicore or FPGA features. Currently, the increasing processing performance of multi/many-core devices, associated with easy-to-use programming models [6,12], made the implementation of prototypes or real communication systems possible. A software-based implementation may not achieve throughput and energy efficiency of ASIC/FPGA implementations. However, it provides flexibility, scalability and enables rapid prototyping for digital communication systems. Nevertheless, achieving high performances is challenging and required algorithm parallelization efforts.

Table 1. Complexity comparison depending on p_Δ value ($p_\omega = 1$)

Method	p_Δ value	Add	Multiply	Memory
FFT	$[1, q[$	$2p_\Delta q\log_2(q)$	$p_\Delta q(\log_2(q)+2)$	$(\frac{p_\Delta - 1}{p_\Delta} + p_\Delta)q$
(distr. memory)	q	$2q^2\log_2(q)$	$q^2(\log_2(q)+2)$	q^2+q+1
FFT	$[1, q[$	*like FFT*	*like FFT*	$\sum_{i=0}^{\log_2(p_\Delta)} \frac{q}{2^i}$
(shared memory)	q	*like FFT*	*like FFT*	$2q-1$
Time Sliding	q	$q(1+q)$	q^2	$2q$
Time Sliding*	q	$q(1+q)$	q	$2q$

Several parallelism levels have been identified and reported. However, the parallelization levels and strategies are influenced by the score method used, the parameters, and the chosen implementation target. The current section first focuses on identifying the system inherent parallelism levels and then details the strategies used for software and hardware targets.

4.1 System Inherent Parallelism

Based on the system description and on Figs. 3 and 4, two coarse grain parallelism levels are unmissable:

1. the Δ-parallelism ($//_\Delta$), related to p_Δ,
2. the ω-parallelism ($//_\omega$), related to p_ω.

The pair p_Δ (in $\{1, 2, 4, \ldots, q\}$) and p_ω (in $\{1, 2, \ldots, 8\}$) impacts on detection performances and on system complexity [5]. Computations performed in $p_\Delta \times p_\omega$ branches are data independent, allowing fully concurrent evaluation.

A third coarse grain parallelism results from the signal oversampling. It is related to the oversampling factor $\mathcal{O}$ (thus noted $\mathcal{O}$-parallelism). It could lead to the execution of $\mathcal{O}$ detectors in parallel, each associated with one oversampling hypothesis.

The three coarse grain parallelism levels enable a large set of setups. Indeed, the overall detection task can be seen as a single task or can be split up to $\mathcal{O} \times p_\Delta \times p_\omega$ parallel sub-tasks.

4.2 Software and Hardware Parallelization

Nowadays, a software platform offers at least two parallelization features being multi-threading (MT) and Single Instruction Multiple Data (SIMD). First, to take advantage of MT, sub-tasks that can be executed in parallel are encapsulated into threads. Threads are especially efficient for coarse grain-level parallelism, involving small interdependent tasks or repetitive processing of different data. This is typically the case for $//_{\mathcal{O}}$, $//_\Delta$ (only with FFT-based SPUs) and seems to be the case for $//_\omega$, regardless of SPU variations. However, as highlighted in the experimentation section, it is not straightforward when subtask complexity is quite low (due to start/join time penalties). At the opposite, SIMD vectorization is involved at the lowest level and requires identical computations that can be applied to multiple data simultaneously. This is exactly the case for $//_\Delta$ in time sliding SPUs. Each correlation kernel can be seen as a parallel computation for each chip of $\mathbf{P}_0$. FFT-based SPU can also benefit from SIMD vectorization using, for instance, the FFTW3 library [10] that takes advantage of such features.

An FPGA device offers more parallelization opportunities than multicore. Indeed, processing elements are designed according to the task specification. It enables to manage easily and efficiently low-level parallelism using sequential, semi-parallel or fully parallel architectures internally. Then, at a higher level, a heterogeneous set of custom processing elements can be allocated to manage sub-tasks concurrently. However, a massive parallelism strategy consisting of duplicating the processing elements is strongly limited by the amount of FPGA resources. Consequently, controllers, memories and logical glue should be added to the system to manage the spatial or temporal reuse of the processing elements. Different SPU trade-off solutions were designed. In all of them, $\mathcal{O} \times p_\omega$ parallel subsystems are allocated for throughput performance reasons as discussed in the experimental section.

5 Implementation Results

The research project aims to implement a complete QCSP communication system. This paper is focused on the transmitter on one hand, and on the first stage of reception, the detection task, on the other hand. Currently, multiple targets are investigated for prototyping and deployment purpose. The Radio-Frequency (RF) related tasks are performed by Ettus Software-defined Radio (SDR) devices. Purposes of the resulting system are to validate the QCSP waveform, and to produce implementation complexity metrics. In this section, transmitter performance levels are first reported. Detection-related performance levels are detailed in a second time.

5.1 Transmitter Implementations

The QCSP waveform improves the communication efficiency by removing the need of a preamble. It is an interesting feature for IoT devices, since those are power constrained low-end systems. A performance evaluation of the emission stack was done. It includes a NB-LDPC encoder, a CCSK mapper, a BPSK modulator and over-modulator as shown in Fig. 1. From now on, sent data consist of a word of 120 bits, or $K = 20$ symbols of $p = 6$ bits, leading to the use of $q = 2^6 = 64$. The NB-LDPC code rate is $R_c = \frac{1}{3}$, which results in an emitted frame of $N = 60$ symbols. These values give an effective code rate $R_{eff} = \frac{1}{32}$ for the communication link, and results in a QCSP frame of 3840 chips.

Table 2. Performance of CCSK software stack on ARM cores.

Target Board/CPU	Clock (GHz)	Length (bits)	Chip rate (MChip/s)	Bit rate (Mbps)	Latency (μs)
Cortex-A53	1.4	120	89	2.78	43
Cortex-A72	1.5	120	200	6.25	19

Table 3. Performance of CCSK hardware stack on Xilinx FPGA.

Target	Clock	Arch.	Chip rate	Bit rate	Latency	Usage		
Name / Serial	(MHz)	(id)	(MChip/s)	(Mbps)	(μs)	FF	LUT	BRAM
Artix 7	100	1	310	9.7	12	870	958	11
xc7a100tcsg324-1		2	86	2.7	88	374	401	12
Spartan 7	100	1	310	9.7	12	870	958	11
xc7s50ftgb196-1		2	96	3.0	80	373	409	12

The software stack was described in C language without leveraging MT nor SIMD optimizations. For evaluation purpose, a set of low-end ARM CPU present on Raspberry Pi boards were selected. Their properties and measured performance levels are provided in Table 2. The achieved information throughput (i.e.

the emission of a frame) on the ARM Cortex-A72 reaches up to 6.25 Mbps. The latency needed to process a frame is equal to 19 μs. Such high throughput on an embedded CPU is due to the low complexity of the encoding and modulation stages. Indeed, the cost of CCSK mapping is low since it is only a rotation, efficiently done by memory remapping and memory copy operations. The throughput and latency parameters measured on the other ARM core are lower, but remains high for the IoT context.

A solution to reduce transmitter power consumption is to design a dedicated circuit to implement the transmitter stack. Consequently, a High Level Synthesis (HLS) friendly C stack description has been designed. Thanks to fast design exploration features provided by the Vivado HLS tool, different hardware implementations have been designed for several low-end FPGAs. The hardware complexity and the latency values are reported in Table 3. Two types of architectures are presented. They were obtained by inserting *pragma* directives, and offer different throughput and complexity trade-offs. The slowest hardware stack (Arch. 2 on Artix 7) is 15× more efficient than the best ARM-based implementation and consumes only 401 LUTs, 374 FFs and 12 BRAMs.

These hardware results highlight the low complexity feature of the transmission process for software and hardware based implementations. The QCSP transmission stack has both low complexity and throughput above 1 Mbps, largely fulfilling, for instance, the LPWAN requirements [1].

5.2 Detector Implementations

The receiver system was implemented on CPU and FPGA targets, however in this case high-end devices are used in line with the much higher computational complexity. Two distinct high-end devices were targeted for the evaluation of the throughput and latency performances of the receiver: an Intel Xeon CPU and a Xilinx Kintex 7 xc7k410tffg900-1, which is the FPGA bundled in the Ettus X310 Universal Software-defined Radio Peripheral (USRP), used as RF receiver. In any case, the $//_{\mathcal{O}}$ is dealt by using $\mathcal{O}$ detectors in parallel, so only the results for one detector are reported in this section.

Software Detector. The communication system reported in Fig. 1 has been described in C++14 language. The software detector has been implemented for the two detailed SPUs variations (FFT and time sliding), for several values of p_{Δ} and p_{ω}, and using two configurations for the time sliding. The high-end multicore system that has been selected for benchmarking is composed of a dual socket Intel Xeon Gold 6148 CPU. It has 256 GB of RAM memory. Each Xeon processor is composed of 20 physical processor cores that share a 28160 KB L3 cache memory. The working frequency of the CPUs is 2.60 GHz but the turbo-boost feature enables cores to run up to 3.70 GHz when the heat dissipation constraint is met. In this system description, the data values are manipulated using single precision elements to avoid precision issues involved by fixed point arithmetic. Consequently, the 12-bit data received from the USRP are converted to float elements.

The parallelization strategy described in the previous section was applied to the C++ source codes. For FFT SPUs, the optimized library FFTW3 [10] is called. This library has CPU detection capabilities, which allows it to execute efficiently the FFTs. It internally uses all applicable CPU features (MT, SIMD, and others). $//_{\Delta}$ is implemented using the OpenMP API, with the FFT shared memory approach described in Sect. 3. The use of OpenMP has been tested for $//_{\omega}$ as well. However, despite our best efforts, it always resulted in a slower implementation.

At the opposite the time sliding correlation method is implemented only thanks to handmade C/C++ codes. The sources are written using floating point values as well, but first as close to the algorithm than possible. Software description was written in a way that takes advantage of GCC auto-vectorization feature, with even some specific software parts finely tuned with intrinsic SIMD instructions.

Measured throughputs for different setups are reported in Table 4. Throughput and latency were measured using the C++14 Chrono API. In Table 4 the MT FFT uses p_{Δ} threads. As it can be seen, the throughput performance depends on the correlation method applied and the selected parameters.

Table 4. Chip rates, bit rates and latencies for different sets of parameters, with $N = 60$, $q = 64$ and an effective rate $R_{eff} = \frac{1}{32}$.

Method	p_{Δ}	p_{ω}	Chip Rate (MChip/s)	Bit Rate (kbps)	Worst Latency (ms)	Best Latency (ms)
FFT	8	4	2.2	69	3.5	1.7
		8	1.3	41	5.8	2.9
	16	4	1.1	34	6.8	3.4
		8	0.68	21	11	5.7
MT FFT	8	4	5.1	160	1.5	0.77
		8	4.1	130	1.9	0.93
	16	4	3.8	120	2	1
		8	3.1	97	2.5	1.2
Time Sliding	64	4	2.4	74	3.2	1.6
		8	1.4	43	5.4	2.7
Time Sliding (optimized)	64	4	3.3	100	2.4	1.2
		8	2.0	63	3.7	1.9

As expected, throughput are lower than transmitter ones, the receiver process being more complex than the transmitter one [14]. On one hand, executed on a single thread, the FFT-based detector is the slowest, achieving at most 2.2 MChip/s, while going as low as 0.64 MChip/s when $p_{\Delta} = 16$ and $p_{\omega} = 8$. On another hand, the throughput is improved up to 5.1 MChip/s by dedicating one physical processor core for each p_Δ filter. Up to 160 kbps of information can be processed as shown in Table 4 (MT FFT). Multi-thread execution provides, in this configuration, a 2.3× speed-up that does not scale with p_Δ value. This performance gap is caused by the overhead incurred by the thread startup and synchronization, due to the OpenMP API.

The time sliding approach, that provides the best detection performances thanks to $p_\Delta = q$, achieves decoding throughput from 1.4 MChip/s when not optimized, and up to 3.3 MChip/s when leveraging vectorization. A key point is that the time sliding method only uses one processor core, and achieving nearly the throughput of the MT FFT nevertheless, consuming at least 16× less CPU resources.

The MT FFT based detection method delivers the highest throughput (5.1 MChip/s) when a large set of cores is available. However, this (p_Δ, p_ω) setup impacts on detection performance. Consequently, the time sliding approach offering higher detection performances and the same high throughput is the best solution. These reception chip rates, which reach a few MChip/s (thus around a hundred kbps), are already similar to those required in, for instance, the Low-Rate Wireless Networks (LRWN) domain [1]. It makes the software receiver a viable solution for the QCSP modulation evaluation.

Software implementations are bound to available CPU features. This is not the case for hardware implementations, detailed in the next sections.

Hardware Implementations. The C++ detector source codes were rewritten in C language using the authorized C synthesis subset [9] supported by Xilinx Vitis HLS 2020 [17]. C descriptions, for both detection approaches, are configurable using *#define* and *#pragma* keywords, controlling the HLS design tool. This highly configurable setup allows testing various architectural variation for different values of p_ω or p_Δ. These C models are synthesized using Vivado HLS for the Xilinx Kintex 7 FPGA available in the X310 USRP.

Table 5. Hardware performance of the detector, clocked at 100 MHz

Method	p_Δ	p_ω	Chip Rate (MChip/s)	Bit Rate (kbps)	Usage FF	LUT	BRAM	DSP
FFT float	8	4	2.1	65	insufficient resources			
FFT fixed 16b	8	4	2.6	81	25726	43206	950	380
		8	2.6	81	51452	94468	1900	760
	16	4	2.6	81	51452	94468	1900	760
		8	2.6	81	102904	188936	3800	1520
Time Sliding float	64	4	9.1	284	144064	80836	79	383
		8	9.1	284	insufficient resources			

Two variants were developed to implement the FFT-based architecture. They were designed with high throughput at all costs as objective, instantiating p_ω processing sub-chains in parallel. They have task pipelining features and parallel processing. The difference is that the first variant use floating-point arithmetic. However, it was requiring too many resources to be put on the chosen target. The second variant has thus been implemented with fixed-point arithmetic, with conservative quantification, to eliminate any risk of performance loss. The FFT models used are derived from work presented in [11]. The floating-point FFT detector reaches 2.1 MChip/s before place and route stage, the same throughput as its software single-core counterparts. The failure to meet the resource constraints was expected, since floating-point arithmetic is known to cause issues on hardware implementations. The fixed-point variant performed slightly better, reaching a throughput of 2.9 MChip/s, this time without exceeding available FPGA resources.

The time-sliding architecture has also been described using pipeline and parallel task execution approaches, as the FFT-based one. However, using the simplification mentioned in the previous section, the computational complexity at $//_\Delta$ is significantly lower. This resulted in a throughput $3\times$ faster than the best FFT implementation, despite the use of floating-point arithmetic. More importantly, its throughput is $1.8\times$ higher than the best software alternative, the MT FFT, which was set up with $p_\Delta = 8$, while the hardware time-sliding detector has a $p_\Delta = 64$, resulting in much higher detection performances. Unfortunately, for $p_\omega = 8$, it failed to meet the resource constraints. To solve this issue and to improve further the architecture efficiency, a fixed-point implementation of the time sliding detector architecture is planned.

6 Conclusion

The paper demonstrated that the QCSP emission and detection tasks can be implemented in real time, allowing reliable transmissions at low SNR without the need of a preamble. The resulting implementations reached throughput from a few dozen kbps to a few hundred kbps, compatible with the LPWAN context, on multicore and FPGA targets. They are especially suitable in a wireless sensor network scenario, with low-end sensor nodes sending data to a higher-end server node. Aside from the score function simplification, this paper also established the time sliding approach as the best to implement the QCSP detection, and detailed several ways to implement it. In the future, a fixed-point version of the time-sliding hardware implementation will be developed to reduce hardware cost and improve the throughput. In parallel, the remaining stages of the QCSP receiver will be implemented to have complete communication system.

References

1. IEEE Std 802.15.4-2020: IEEE Standard for Low-Rate Wireless Networks (2020)
2. Abassi, O., Conde-Canencia, L., Mansour, M., Boutillon, E.: Non-binary low-density parity-check coded cyclic code-shift keying. In: Proceedings of WCNC, Shanghai, China, April 2013. https://doi.org/10.1109/WCNC.2013.6555196
3. Azari, A., al.: Grant-free radio access for short-packet communications over 5G networks. In: Proceedings of GLOBECOM (2017)
4. Bloessl, B., Dressler, F.: mSync: physical layer frame synchronization without preamble symbols. IEEE Trans. Mob. Comput. **17**(10), 2321–2333 (2018)
5. Camille, M., Kassem, S., Le Gal, B., Boutillon, E.: Time sliding window for the detection of CCSK frames. In: Proceedings of SiPS. IEEE (2021)
6. Checko, A., et al.: Cloud RAN for mobile networks - a technology overview. IEEE Commun. Surv. Tutorials **17**(1), 405–426 (2015)
7. Dillard, G., et al.: Cyclic code shift keying: a low probability of intercept communication technique. IEEE Trans. Aerosp. Electron. Syst. **39**(3), 786–798 (2003)
8. Durisi, G., et al.: Toward massive, ultrareliable, and low-latency wireless communication with short packets. Proc. IEEE **104**(9), 1711–1726 (2016)
9. Fingeroff, M.: High-Level Synthesis Blue Book. Xlibris Corporation, Bloomington (2010)
10. Frigo, M., Johnson, S.: The design and implementation of FFTW3. Proc. IEEE **93**(2), 216–231 (2005)
11. Kastner, R., Matai, J., Neuendorffer, S.: Parallel Programming for FPGAs. ArXiv e-prints (2018)
12. Mavromoustakis, C.X., Mastorakis, G., Dobre, C. (eds.): Advances in Mobile Cloud Computing and Big Data in the 5G Era. SBD, vol. 22. Springer, Cham (2017). https://doi.org/10.1007/978-3-319-45145-9
13. Polyanskiy, Y.: Asynchronous communication: exact synchronization, universality, and dispersion. IEEE Trans. Inf. Theory **59**(3), 1256–1270 (2013)
14. Saied, K.: Quasi-Cyclic Short Packet (QCSP) Transmission for IoT. Theses, Université Bretagne Sud, March 2022
15. Saied, K., Ghouwayel, A., Boutillon, E.: Time-synchronization of CCSK short frames. In: Proceedings of WiMob (2021)
16. Walk, P., et al.: MOCZ for blind short-packet communication: practical aspects. IEEE Trans. Wireless Commun. **19**(10), 6675–6692 (2020)
17. Xilinx: Vitis High-Level Synthesis User Guide UG1399 (v2021.1), June 2021

Towards Lightweight Deep-Learning Techniques

Dynamic Pruning for Parsimonious CNN Inference on Embedded Systems

Paola Busia[1(✉)], Ilias Theodorakopoulos[2], Vasileios Pothos[2], Nikos Fragoulis[2], and Paolo Meloni[1]

[1] Università degli Studi di Cagliari, Cagliari, Italy
paola.busia@unica.it
[2] Irida Labs, Magoula, Greece

Abstract. As a consequence of the current edge-processing trend, Convolutional Neural Networks (CNNs) deployment has spread to a rich landscape of devices, highlighting the need to reduce the algorithm's complexity and exploit hardware-aided computing, as two prospective ways to improve performance on resource-constrained embedded systems. In this work, we refer to a compression method reducing a CNN computational workload based on the complexity of the data to be processed, by pruning unnecessary connections at runtime. To evaluate its efficiency when applied on edge processing platforms, we consider a keyword spotting (KWS) task executing on SensorTile, a low-power microcontroller platform by ST, and an image recognition task running on NEURAghe, an FPGA-based inference accelerator. In the first case, we obtained a 51% average reduction of the computing workload, resulting in up to 44% inference speedup, and 15% energy-saving, while in the latter, a 36% speedup is achieved, thanks to a 44% workload reduction.

Keywords: Convolutional Neural Networks · Pruning · Hardware acceleration

1 Introduction

Convolutional Neural Networks (CNNs) have reached outstanding levels of accuracy [1], favoring their success in multiple application fields, from natural language processing, to image classification, and object detection. A turning point was represented by the design of deeper and complex architectures [2], having pushing requirements in terms of storage and computing capabilities. Their deployment on edge resource-constrained systems, encouraged by bandwidth, security, and privacy concerns, poses many challenges and has been a prolific field of research. On the one hand, more efficient hardware architectures, specifically targeting neural networks, have been designed. Industry and academia have proposed multiple dedicated processors and accelerators [3–7] and embedded GPUs [8], and heterogeneous computing systems exploiting FPGAs and All-Programmable-SoCs to combine parallelism and flexibility [9,10]. On the

K. Desnos and S. Pertuz (Eds.): DASIP 2022, LNCS 13425, pp. 45–56, 2022.
https://doi.org/10.1007/978-3-031-12748-9_4

other hand, optimized software libraries, specifically targeting a class of devices, have been developed and released, such as CU-DNN for NVIDIA platforms [11], CMSIS [12], targeting class-M ARM low-power microprocessors, and ARM-NN [13], targeting more high-end architectures. Moreover, multiple approaches have been focusing on the simplification of the computing model, to reduce the footprint, or the computing workload of CNNs [14,15].

In this work, we mean to leverage the combination of such approaches. We take as a reference a technique, proposed in [16], performing an online pruning of a CNN's connections, reducing the computational load associated with convolutional layers, based on run-time processing of the input data. We test its implementation on a resource-constrained commercial platform based on ARM Cortex-M, the ST SensorTile, considering a CNN for a keyword spotting (KWS) task, and evaluating the obtained improvement in terms of execution time and power consumption. We also present an image recognition use case, exploiting a custom network architecture built for CIFAR-10 [17], to evaluate the feasibility of dynamic pruning on a hardware-assisted computing platform, considering as a target NEURAghe, an FPGA-based inference accelerator.

2 Related Work

The efficient execution of CNNs on resource-constrained systems requires careful optimization, both of the computational workload and of the number of accesses to the off-chip memory. The community has addressed this matter by either designing shallower and optimized network architectures [18,19], or by developing several compression techniques, reducing the number of network's parameters or the precision of their representation [20]. In Table 1 we list some of the most recent works on network compression. For each one, we define in Column 1 the dominant compression method resulting in most of the reported advantages, while in Column 2 we report whether the compression strategy is *static* or *dynamic*, thus evaluated at runtime based on the complexity of the input to be classified. In Column 3 we define the granularity and structure level of the pruning action. In Column 4 we report the performance metrics considered to evaluate and refine the CNN architecture, and finally in Column 5 we list the hardware architectures considered for the analysis.

In [21], the authors present an hybrid neural network, combining the advantages of Strassen representation for matrix multiplications [24] and of Bonsai decision trees [25]. Their proposed compression method exploits ternary representation for most of the weights in convolutional layers, keeping only a few full precision weights, which can be further quantized to 16 or 8-bit precision. The ST-HybridNet reaches 94.71% accuracy, using 2.4 MOPS and requiring a 41.8 kB memory footprint. The compression method exploited in this work is static, and its advantages are mainly due to compression through quantization rather than connections pruning, and the possibility to replace most of the resource-hungry multiplications with additions. Thus, it can be considered as an orthogonal technique to classical pruning methods, and especially to dynamic

Table 1. Comparison with state of the art works on CNN compression.

Work	Compression Strategy	Static/ Dynamic	Structure Level	Performance Metric	Hardware Platform
[21]	Transformation, Quantization	Static	not applicable	Accuracy, OPS, Footprint	✗
[22]	Pruning, Quantization	Dynamic	Group of channels	Accuracy, OPS, Power	ASIC, GPU, CPU
[23]	Pruning	Dynamic	Group of channels	Accuracy, OPS, Power	GPU
this work	Pruning	Dynamic	Single channel	Accuracy, inference time, Power	FPGA, μC

ones. Furthermore, the advantages of compression are only indirectly analyzed in terms of OPS and footprint reduction, which cannot always be translated into performance improvements, depending on the flexibility of the target library.

In [22], the authors explore channel gating as a dynamic pruning method, to reduce at runtime the network's complexity based on the input's content. Given a baseline network architecture, two different paths are identified, a *base path*, and a *conditional path*. The first one is always computed, and such a partial sum is exploited as an activation rule for the *conditional path*, which is either skipped or selectively executed. In the case of skipped computations, the partial sum is used as an approximation of the final value, thus the workload reduction is not inherited by the successive layers. Thus, in the resulting channel gating networks (CGNets) a structured pruning is enforced, which allows for efficient inference, even on a hardware accelerator. The authors report a 2.8× FLOPS reduction on ResNet-18, resulting in a 2.3× inference speedup on their CGNet accelerator.

In [23], the authors present a dynamic pruning strategy based on reinforcement learning, exploiting a decision network to evaluate the pruning actions on the convolutional layers of the network. For each layer, the output features are grouped into a certain number, k, of sets, and the decision establishes how many of such ordered sets are to be evaluated based on the desired trade-off between performance and accuracy. Thus, it is still possible to define a *base* and a *conditional path*, and the enforced pruning can be defined as structured.

In this work, we consider a dynamic pruning technique [16], which can be exploited concurrently with other compression methods, to further reduce the workload of an efficient network based on the content of the specific input to be processed. Compared to [22,23], we focus on a less structured pruning strategy, where the activation rule is applied independently to the single output features of a convolutional layer, and all the combinations can be in theory obtained. The compression effect also impacts the following layers, resulting in fewer valid input features to be computed in the successive convolutions. The work we reference [16] represents an extreme case of dynamic pruning, thus we mean to evaluate whether the resulting OPS reduction can still produce performance improvement. We consider an efficient state-of-the-art library, as CMSIS-NN, enabling it to support the selective evaluation of convolutional kernels, and finally evaluate the advantages of the dynamic pruning on on-hardware direct measurements of inference time and energy consumption. We also consider a convolution accelerator implemented on FPGA and evaluate how the workload reduction impacts its performance.

In the following, we show that significant performance improvement can be achieved by introducing the required support in the hardware architecture or software library performing the convolutions. Our purpose is to:

- test the pruning method in [16] on two use-cases, KWS and image recognition;
- evaluate its effect on two reference hardware platforms, the ST SensorTile and NEURAGHE;
- present a method to estimate the Parsimonious Inference's (PI) impact on power consumption.

3 Reference Methodology

The common inspiration for pruning methods is CNN computations are often redundant, and some of them can be skipped with little effect on the accuracy. [16] focuses on convolutional layers, which represent the main source of workload for several CNN architectures, and aims at adapting the complexity of the network to the particular item to be classified. The approach exploits a specialized training procedure, where the network model is changed into one enforcing PI through dynamic pruning. As shown in Fig. 1, the network architecture is distorted through the insertion of a dedicated software module, the Learning Kernel Activation Module (LKAM), which can be associated with one or more convolutional layers along the network. The LKAM reproduces a simple network model, consisting of a 1×1 convolution, average pooling, and a sigmoid activation, followed by a threshold step. During the inference execution, this lightweight processing is applied to the convolutional layer's input features, resulting in a set of activation flags, provided as an additional input. In detail, given a layer with OF output features, the LKAM computes OF activation flags. At runtime, convolution is evaluated only on the *active* output features, thus skipping the computations associated with particular sets of weights. Since the LKAM output is data-dependent, different levels of deactivation can be obtained for different input items. The computation savings also involve the following layer, which will receive a reduced number of valid input features. The LKAM parameters are learned during the training procedure, aiming at preserving the network's accuracy while maximizing the sparsity of the activation flags. To that end, the

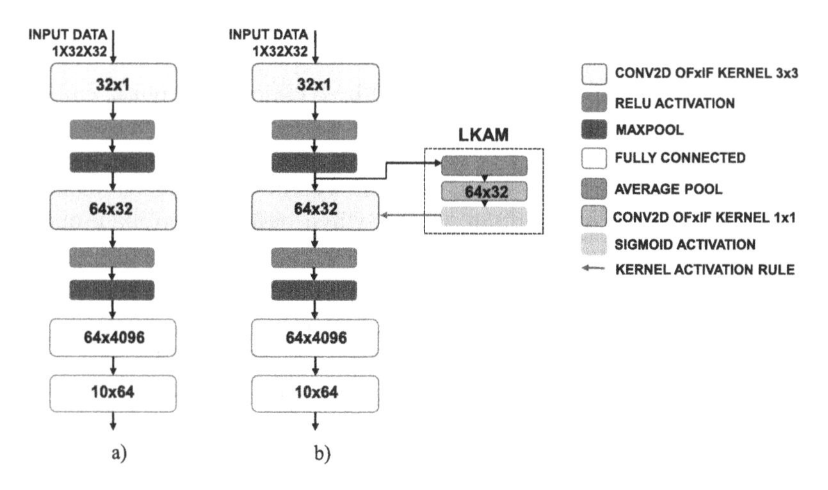

Fig. 1. a) Network architecture for the KWS use-case; b) Parsimonious model obtained through retraining with LKAMs.

chosen loss function, $L_t(w, b)$, needs to be modified through an additional term, indicated as $L_{aug}(sw)$, to consider the LKAM output.

$$L(w, b, sw) = L_{\mathrm{t}}(w, b) + L_{\mathrm{aug}}(sw)$$

Such new term is defined as:

$$L_{aug}(sw) = \frac{G_{\mathrm{i}}}{2m} \sum_{i} |sw_i|$$

where sw represents the LKAM output before the threshold is applied, m is the length of such vector and G_i is a gain factor, that can be tuned independently for each layer of the network. The tool output is a set of new models, integrating different numbers of trained LKAMs in different positions. When a target accuracy is set, the selected model is the one with the highest deactivation percentage within the target accuracy.

When trying to exploit this technique on hardware-aided or parallel computing devices, the main challenge is posed by the need to selectively perform computations, and keep track of the deactivated features when retrieving the appropriate kernels to evaluate across the various layers of the network. A possible drawback is represented by the additional storage space required by the LKAM's parameters. Loss of precision, due to the need for data quantization, is another issue to be considered in both the implementations evaluated in this paper.

4 Reference Computing Platforms

We present in the following the main features of the platforms considered in this paper, the ST SensorTile and NEURAGHE, and the modifications we introduced to support PI.

4.1 SensorTile

The SensorTile is an IoT module, developed by STMicroelectronics, embedding an 80 MHz ARM Cortex-M4 32-bit low-power micro-controller. The system architecture exploits a Real-Time lightweight Operating System (RTOS), supporting multi-threading and scheduling of the different application tasks on defined timings. To reduce power consumption, it can switch between two main operating states, *run mode* and *sleep mode*, exploited whenever possible in our application, through a specific *idle* task that is entered every time none is pending.

Support for PI. Given the real-time constraints of the KWS task on SensorTile, we exploited CMSIS, a library specifically targeting Cortex-M Processing Cores, including several NN utilities and designed to maximize performance [12]. To obtain the results presented in the last section, we used the basic version of the 8-bit square convolution function provided by the library, *"arm_convolve_HWC_q7_basic"* and customized it to make it able to receive and use the deactivation information produced by the LKAMs. As a first attempt at supporting PI, after scanning the activation flags we split the convolution execution into separate layers, of width given by the number of consecutive active output features.

As shown in Fig. 2, where the execution time of layer Conv2 in Table 2 is plotted as a function of the kernels' deactivation percentage, such a solution is not very efficient, preventing the processor to take advantage of the optimized sequence of operations. As an alternative, we acted on the function itself, to introduce the selective evaluation of computational units, while preserving the computational efficiency. We replaced the weight tensor with an array of addresses, each pointing to the active filtering kernels. During convolution execution, the read pointer of weights is assigned a new value from the next location of the addresses tensor. As can be derived from Fig. 2, introducing kernel deactivation inside a knowingly optimized function allows obtaining a linear speedup with the deactivation percentage.

4.2 NEURAghe

NEURAghe is a CNN inference accelerator that can be ported with different parameters on different FPGA devices [10]. The results disclosed in this work come from its implementation on a Xilinx Z-7020 SoC mounted on a Zedboard by Digilent. It exploits a Convolution Engine (CE) embedding a matrix of multipliers, and a programmable micro-controller, efficiently scheduling convolutions and data transfers towards the local storage space accessed by the CE, dedicated to the convolutional weights and activations. In this work, we only refer to the hardware acceleration of convolutional layers, with kernel size 3×3 or 5×5. When receiving the offload command, the micro-controller is provided with the layer's parameters, and the memory addresses from which to read the network's parameters and write the computed results. According to the internal structure of the CE and the size of the local storage space, the micro-controller groups the layer's input and output features, and handles a task-level pipeline, made

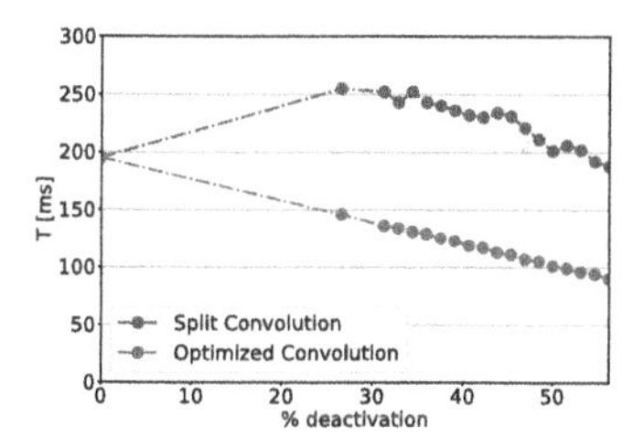

Fig. 2. Execution time of Conv2 layer in the KWS network model, considering different percentages of deactivated kernels and exploiting the two solutions developed to support kernel deactivation.

up of three main stages: 1) *load* of weights and input features; 2) setup and *run* of the CE; 3) *store* of the computed results. The disclosed results refer to a CE implementation embedding a single SoP module, processing 2 output pixels per cycle, and performing three convolutions with 3×3 filtering kernels, or one with 5×5 kernels, in 16-bit fixed-point precision.

Support for PI. To efficiently support PI, the hardware/software architecture needs to selectively load only the sets of weights corresponding to active kernels, while exploiting at best the hardware resources, by setting up the engine with the optimal number of inputs and weights, from the scheduling point of view. First, the layer's description provided to the micro-controller is enriched with two fields, carrying kernel activation information, and consisting of two arrays of flags, storing the output of the LKAM associated with the current layer, and with the preceding one. Given such information, the micro-controller needs to program the CE to only perform the computations corresponding to a valid input feature and active output feature. Furthermore, in the baseline architecture, weight transfers are handled in batches, while PI requires treating each kernel independently, to freely discard those that have been deactivated. To enable this, we changed weight transfer granularity and added the memory-mapped programmable registers needed to control the number of elements to be expected per transfer, and the position of the element that should be interpreted as bias. We also introduced some changes in the middleware executed by the micro-controller, to evaluate the activation flags before the data transfers are programmed. An outer loop scans the flags associated with the input and output features: only when both flags are set to true the address is evaluated, and the transfer is programmed. Finally, we introduced two inner loops on the activation flags, to keep track of the number of programmed transfers, until the necessary number of features, required to run the accelerator at full speed, has been reached.

5 Experimental Results

5.1 KWS on SensorTile

To evaluate the advantages of PI in terms of power consumption, we selected a KWS use-case, deployed on ST SensorTile. Classification is performed through a

Table 2. Architectural model of the reference CNN for the KWS task.

Layer		
Conv1; Convolution	Input Size = 32 × 32 Input Features = 1	Kernel Size = 3 × 3 Output Features = 32
Conv2; Convolution	Input Size = 16 × 16 Input Features = 32	Kernel Size = 3 × 3 Output Features = 64
Fc1; Fully Connected	Input Size = 64 × 8 × 8	Output Size = 64
Fc2; Fully Connected	Input Size = 64	Output Size = 10

Table 3. Accuracy of the baseline and PI version of the CNN for KWS.

	Accuracy
Baseline	94,43%
PI	90,54%

Table 4. On-hardware measurements of power consumption and energy contributions of the examined tasks, performed on the ST SensorTile.

Idle	Power: P_{idle} = 28,78 mW
Audio Processing	Energy Contribution: E_{pr} = 3,48 mJ
CNN Classifier	Energy Contribution: E_{cnn} = 8,67 mJ

simple CNN model architecture, trained on the Speech Commands dataset [26], and whose structure is described in Table 2. We refer to the model enriched with an LKAM associated with Conv2 as the PI model, resulting in the accuracy drop reported in Table 3. Figure 3a reports the different percentages of deactivation obtained over the dataset, showing an average deactivation of 51% of the convolutional kernels in Conv2, and evaluated on a 16-bit implementation. To assess how power consumption is affected, we refer to an 8-bit implementation, running on the reference platform. We exploited the CMSIS library [12], introducing the modifications described in Sect. 4.1 to support the selective evaluation of kernels. Figure 3b reports the inference time as a percentage of the baseline execution time, considering the kernels activation percentages in Fig. 3a. To translate processing time savings into a reduction of power consumption, we considered a simple application model, involving three main tasks:

- *Get Data*, performing data acquisition with the desired sampling frequency;
- *Audio Processing*, evaluating the Mel-spectrogram of the sampled audio, provided as input to the CNN;
- *CNN Classifier*, executing the network model for recognition.

The audio processing is performed by evaluating 32 Mel features through 32 temporal frames, covering 1s of sampled audio. Considering real-time execution,

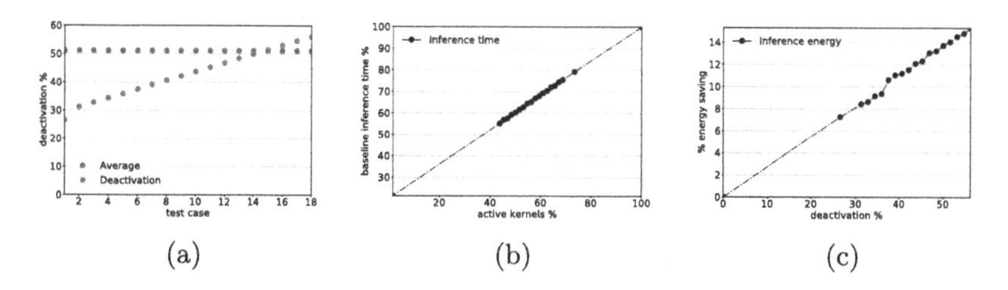

Fig. 3. a)Kernel deactivation percentage through the Speech Commands dataset, and average deactivation; b)Percentage execution time over the baseline model execution time, for different kernel activation levels; c)Percentage energy saving over the baseline model power consumption, for different percentages of kernel deactivation.

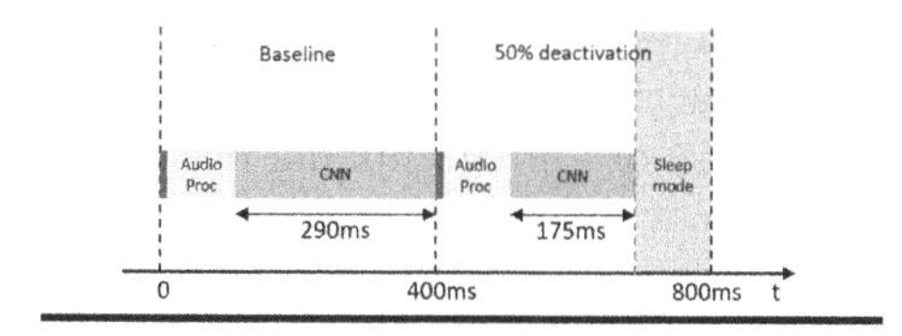

Fig. 4. Real-time inference cycle, considering the baseline model and a case of 50% deactivation in Conv2, allowing to save power by entering the sleep mode.

processing and classification need to be performed periodically. Since the audio processing time was evaluated to be 104 ms at an 80 MHz working frequency, and the worst-case inference time, required by the baseline model without deactivation, is 290 ms, we considered a fixed execution period of 400 ms. When no software task is executing, the RTOS switches the operating state from *run* to *sleep* mode. As shown in Fig. 4, the idle time grows with the percentage of kernel deactivation produced by the input. The effect on the system's power consumption can be estimated through a simple model [27]:

$$P = P_{\text{idle}} + E_{\text{gd}} \times f_{\text{gd}} + (E_{\text{pr}} + E_{\text{cnn}}) \times f_{\text{pr}}$$

where f_{pr} is the execution frequency of processing and classification, while with E_{pr} and E_{cnn} we refer to the energy contribution of the different tasks, that add up to the idle power consumption of the system. Table 4 reports the idle power consumption and the energy contribution of the considered tasks, evaluated by measuring with an oscilloscope the current absorbed by the device in different working conditions. Figure 3c reports the energy consumption, as a percentage of E_{cnn} in Table 4, based on the fixed deactivation patterns in Fig. 3a, and referring to a loop execution of the CNN classifier. The results in this section show that the examined pruning method can produce remarkable hardware performance improvements, despite the high level of granularity.

5.2 CIFAR-10 on NEURAGHE

Here we explore the feasibility of applying PI on a hardware assisted architecture, considering an image recognition task executing on NEURAGHE, and exploiting a simple network model, LeNet, trained on CIFAR-10 [17]. The network structure is summarized in Table 5. The PI version of the model was obtained by associating LKAMs to the second and third convolutional layers, with an accuracy drop of around 3%. Table 6 reports three test cases, producing different deactivation patterns, and resulting in a different inference speedup, evaluated considering: 1) software execution on ARM Cortex A9 processor (667 MHz); 2) hardware execution exploiting NEURAGHE's CE. As shown in the Table, in the case of software execution the additional overhead introduced by dynamic pruning is completely compensated by the reduction of computations to be performed, producing significant advantages in the overall execution. When convolutions are handled on FPGA, kernel deactivation still results in performance improvement, but it is less effective, because of a programming overhead, whose burden does not depend on the size of the workload.

Table 5. Architectural model of the considered LeNet for CIFAR-10.

Layer		
Conv1; Convolution	Input Size = 32×32 Input Features = 3	Kernel Size = 5×5 Output Features = 32
Conv2; Convolution	Input Size = 16×16 Input Features = 32	Kernel Size = 5×5 Output Features = 32
Conv3; Convolution	Input Size = 8×8 Input Features = 32	Kernel Size = 5×5 Output Features = 64
Fc1; Fully Connected	Input Size = $64 \times 4 \times 4$	Output Size = 64
Fc2; Fully Connected	Input Size = 64	Output Size = 10

Table 6. Kernel deactivation percentage in the three test cases, and PI execution speedup on ARM Cortex A9 (667 MHz) and NEURAGHE Convolution Engine on Xilinx Zynq Z-7020.

	Test 1			**Test 2**			**Test 3**		
	Conv 2	*Conv 3*	*Tot*	*Conv 2*	*Conv 3*	*Tot*	*Conv 2*	*Conv 3*	*Tot*
Deactivation	31.3%	31.3%	30.3%	43.8%	48.1%	45.3%	40.6%	40.6%	39.4%
ARM speedup	30%	27%	29%	43%	44%	43%	40%	36%	39%
CE speedup	19%	24%	22%	33%	36%	35%	31%	30%	30%

6 Conclusion

In this work, we presented the implementation of dynamic neural network pruning through data-driven kernel deactivation on two resource-constrained platforms, exploiting different computing units, SensorTile and NEURAghe. We referred to two common application fields of CNNs, such as image recognition and KWS, and considered custom network architectures. Referring to common datasets, we found that the method allows an average deactivation of 51% of the convolutional kernels in the KWS task. The experimental results show that the reduced computational load creates the possibility to reduce the system's power consumption, up to 15% of energy-saving, corresponding to a 44% speedup. The data on NEURAghe implementation show it is possible to exploit dynamic deactivation even when adopting FPGA acceleration, although with less effective improvements. In this case, a maximum 36% speedup due to a 44% deactivation is obtained.

References

1. LeCun, Y., Bengio, Y., Hinton, G.: Deep learning. Nature **521**, 436–444 (2015)
2. He, K., Zhang, X., Ren, S., Sun, J.: Deep residual learning for image recognition (2015)
3. Jouppi, N.P., et al.: In-datacenter performance analysis of a tensor processing unit. In: ISCA 2017: Proceedings of the 44th Annual International Symposium on Computer Architecture, pp. 1–12, June 2017. https://doi.org/10.1145/3079856.3080246
4. Azarkhish, E., Rossi, D., Loi, I., Benini, L.: Neurostream: scalable and energy efficient deep learning with smart memory cubes. IEEE Trans. Parallel Distrib. Syst. **22**(2), 420–434 (2018)
5. Desoli, G., et al.: 14.1 a 2.9TOPS/W deep convolutional neural network SoC in FD-SOI 28 nm for intelligent embedded systems. In: 2017 IEEE International Solid-State Circuits Conference (ISSCC), pp. 238–239 (2017)
6. Movidius: Movidius neural compute stick: accelerate deep learning development at the edge (2020). https://developer.movidius.com/
7. Chen, Y.-H., Emer, J., Sze, V.: Eyeriss: a spatial architecture for energy-efficient dataflow for convolutional neural networks. In: 2016 ACM/IEEE 43rd Annual International Symposium on Computer Architecture (ISCA), pp. 367–379 (2016)
8. NVIDIA: Nvidia deep learning accelerator (2020). https://developer.nvidia.com/embedded/buy/tegra-k1-processor
9. Blott, M., Preusser, T., Fraser, N., Gambardella, G., O'Brien, K., Umuroglu, Y.: FINN-R: an end-to-end deep-learning framework for fast exploration of quantized neural networks. ACM Trans. Reconfigurable Technol. Syst. (TRETS) (2018). https://doi.org/10.1145/3242897
10. Meloni, P., et al.: NEURAghe: exploiting CPU-FPGA synergies for efficient and flexible CNN inference acceleration on Zynq SoCs. ACM Trans. Reconfigurable Technol. Syst. (TRETS) (2018). https://doi.org/10.1145/3284357
11. NVIDIA: cuDNN (2020). https://developer.nvidia.com/cudnn
12. Lai, L., Suda, N., Chandra, V.: CMSIS-NN: efficient neural network kernels for Arm Cortex-M CPUs. CoRR, abs/1801.06601 (2018). http://arxiv.org/abs/1801.06601

13. ARM-NN (2020). https://www.arm.com/products/silicon-ip-cpu/machine-learning/arm-nn
14. Han, S., et al.: EIE: efficient inference engine on compressed deep neural network. In: ISCA 2016: Proceedings of the 43rd International Symposium on Computer Architecture, pp. 243–254, June 2016. https://doi.org/10.1109/ISCA.2016.30
15. Han, S., et al.: ESE: efficient speech recognition engine with sparse LSTM on FPGA. In: FPGA 2017: Proceedings of the 2017 ACM/SIGDA International Symposium on Field-Programmable Gate Arrays, pp. 75–84, February 2017. https://doi.org/10.1145/3020078.3021745
16. Theodorakopoulos, I., Pothos, V., Kastaniotis, D., Fragoulis, N.: Parsimonious inference on convolutional neural networks: learning and applying on-line kernel activation rules. CoRR, abs/1701.05221 (2017). https://arxiv.org/abs/1701.05221
17. Krizhevsky, A.: Learning multiple layers of features from tiny images (2009). https://www.cs.toronto.edu/~kriz/learning-features-2009-TR.pdf
18. Iandola, F.N., et al.: SqueezeNet: AlexNet-Level accuracy with 50x fewer parameters and <0.5 mb model size. CoRR, abs/1602.07360 (2016). http://arxiv.org/abs/1602.07360
19. Zhang, Y., Suda, N., Lai, L., Chandra, V.: Hello edge: keyword spotting on microcontrollers. CoRR, arXiv:1711.07128 (2017)
20. Han, S., Mao, H., Dally, W.J.: Deep compression: compressing deep neural networks with pruning, trained quantization and Huffman coding. In: International Conference on Learning Representations 2016, October 2015. https://arxiv.org/abs/1510.00149
21. Gope, D., Dasika, G., Mattina, M.: Ternary hybrid neural-tree networks for highly constrained IoT applications (2019)
22. Hua, W., Zhou, Y., De Sa, C., Zhang, Z., Suh, G.E.: Boosting the performance of CNN accelerators with dynamic fine-grained channel gating. In: Proceedings of the 52nd Annual IEEE/ACM International Symposium on Microarchitecture, ser. MICRO 52, pp. 139–150. Association for Computing Machinery, New York (2019). https://doi.org/10.1145/3352460.3358283
23. Lin, J., Rao, Y., Lu, J., Zhou, J.: Runtime neural pruning. In: Advances in Neural Information Processing Systems, vol. 30. Curran Associates, Inc. (2017). https://proceedings.neurips.cc/paper/2017/file/a51fb975227d6640e4fe47854476d133-Paper.pdf
24. Tschannen, M., Khanna, A., Anandkumar, A.: StrassenNets: deep learning with a multiplication budget. In: Dy, J., Krause, A. (eds.) Proceedings of the 35th International Conference on Machine Learning, ser. Proceedings of Machine Learning Research, vol. 80, pp. 4985–4994. PMLR, 10–15 July 2018. https://proceedings.mlr.press/v80/tschannen18a.html
25. Kumar, A., Goyal, S., Varma, M.: Resource-efficient machine learning in 2 KB RAM for the internet of things. In: Precup, D., Teh, Y.W. (eds.) Proceedings of the 34th International Conference on Machine Learning, ser. Proceedings of Machine Learning Research, vol. 70, pp. 1935–1944. PMLR, 06–11 August 2017. https://proceedings.mlr.press/v70/kumar17a.html
26. Warden, P.: Speech commands: a dataset for limited-vocabulary speech recognition. CoRR, arXiv:1804.03209 (2018)
27. Scrugli, M.A., Loi, D., Raffo, L., Meloni, P.: A runtime-adaptive cognitive IoT node for healthcare monitoring. In: Proceedings of the 16th Conference on Computing Frontiers (CF 2019), pp. 350–357, April 2019. https://doi.org/10.1145/3310273.3323160

DL-CapsNet: A Deep and Light Capsule Network

Pouya Shiri(✉) and Amirali Baniasadi

University of Victoria, Victoria, BC, Canada
{pouyashiri,amiralib}@uvic.ca

Abstract. Capsule Network (CapsNet) is among the promising classifiers and a possible successor of the classifiers built based on Convolutional Neural Network (CNN). CapsNet is more accurate than CNNs in detecting images with overlapping categories and those with applied affine transformations. In this work, we propose a deep variant of CapsNet consisting of several capsule layers. In addition, we design the Capsule Summarization layer to reduce the complexity by reducing the number of parameters. DL-CapsNet, while being highly accurate, employs a small number of parameters and delivers faster training and inference. DL-CapsNet can process complex datasets with a high number of categories.

Keywords: Capsule Networks · Deep CapsNet · Fast CapsNet

1 Introduction

Sabour et al. introduced Capsule Network (CapsNet) [1] as the new generation of classifiers with several advantages over traditional Convolutional Neural Networks (CNNs). CapsNet is more robust to applying affine transformations and detects images with overlapping categories easier than CNNs. CapsNet offers competitive accuracy showing promising results on small-scale datasets such as MNIST [2] and Fashion-MNIST [3]. On more complex datasets such as CIFAR-10 and CIFAR-100 [4], however, the results are not as competitive. There have been several works aiming at facilitating supporting datasets with a high number of categories.

The basic computational unit of CapsNet is referred to as a capsule (a vector of neurons). CapsNet consists of a simple feature extractor including two convolutional layers. The extracted features are then reshaped to vectors. These vectors are multiplied by multiple matrices to produce the first level of capsules referred to as Primary Capsules (PCs). The next layer of capsules (output capsules) are generated out of PCs using an iterative algorithm called Dynamic Routing (DR). In DR, all input capsules contribute to all output capsules but with different weights. The output capsules are used for the classification. CapsNet employs a simple decoder consisting of Fully-Connected (FC) layers to regularize training by adding the reconstruction term to the loss function.

K. Desnos and S. Pertuz (Eds.): DASIP 2022, LNCS 13425, pp. 57–68, 2022.
https://doi.org/10.1007/978-3-031-12748-9_5

We propose a deep network to add support for more complex datasets. Making networks deeper (stacking up layers) results in high generalization, and hence is common. However, deeper networks have higher number of parameters (trainable weights). This is critical as it affects the computation cost and the resources required. Therefore, we take further measures to reduce complexity by developing a mechanism to reduce the number of capsules. This is achieved by replacing multiple capsules with only a few using a summarization mechanism.

We introduce the Capsule Summarization (CapsSum) layer and summarize the generated capsules into only a few. The reduction in the number of capsules reduces the number of parameters and speeds up the network.

Deeper networks deeper usually have higher representation ability. Therefore, we started by making the network deeper. Stacking several fully-connected capsule layers with DR inferring the capsules from one layer to the next is computationally expensive. Moreover, using the DR algorithm multiple times, leads to poor learning in the intermediate layers [5]. However, there are different DR alternatives, and as we show, employing a three-dimensional 3DR algorithm [6]) is a reasonable alternative.

We introduce Multi-Level Capsule Extractor (MLCE), that uses the 3DR algorithm twice and includes two CapsSum layers. MLCE takes capsules as input, and outputs a combination of high-level and low-level capsules. MLCE consists of convolutional layers, CapsSum, and 3DR. It generates fewer capsules in two levels that provide a robust part-to-whole representation and reduces the total number of parameters in the network.

Networks based on CapsNet usually include a decoder to reconstruct the input images and avoid over-fitting. In order to maintain accuracy, we carefully employ an efficient decoder (class-independent decoder).

In summary, we propose DL-CapsNet as a deep, light and highly accurate variant of CapsNet by using the MLCE module on top of a deep convolutional sub-network, and employing an efficient decoder. Our contributions are:

- We develop a deep network achieving high test accuracy. DL-CapsNet includes several capsule layers, and uses 3DR twice to make a high-level representation of the input images. We achieve 91.23% accuracy for CIFAR-10 using a 7-ensemble model.
- We reduce complexity by introducing the CapsSum layer. CapsSum reduces the number of generated capsules by using a deep structure consisting of several capsule layers. DL-CapsNet contains 6.8M parameters.
- We evaluated the network for CIFAR-10, SVHN, and Fashion-MNIST and achieved state-of-the-art results. In addition, we support more complex datasets. Using a 7-ensemble model for CIFAR-100, we achieve 68.36% accuracy.

The rest of this paper is organized as follows. The related works are presented in Sect. 2. Section 3 reports the background. DL-CapsNet is presented in Sect. 4. Experiments and results are presented in Sect. 5. The paper is concluded in Sect. 6.

2 Related Works

Several studies have improved the CapsNet networks. Yang et al. [7] proposed RS-CapsNet. RS-CapsNet integrates Res2Net blocks to extract features in multiple scales. It also uses the Squeeze-and-Excitation (SE) block to emphasize more useful features. In order to enhance the representation ability and reduce the number of capsules, RS-CapsNet uses a linear combination of capsules.

Huang et al. proposed DA-CapsNet [8], which uses the attention mechanism after the convolution layers and also after the primary capsules layer. DA-CapsNet is highly accurate for SVHN, CIFAR10, FashionMNIST, smallNORB, and COIL-20 datasets. DA-CapsNet outperforms CapsNet in image reconstruction.

Shiri et al. [9] proposed Quick-CapsNet (QCN) modifying the low-level feature extraction, resulting in only a few capsules. The reduction of PCs results in a significant speedup at the cost of marginal loss of accuracy. QCN uses an optimized decoder to improve the network generalization.

Shiri et al. [10] proposed CFC-CapsNet. CFC-CapsNet used a new layer for creating PCs out of the extracted features. This layer results in fewer number of capsules while improving accuracy. The reduction of number of capsules, reduces the overall number of parameters as well as speeding up the network.

Deliege et al. [11] proposed HitNet which replaces a layer with a Hit-or-Miss layer. Capsules in this layer are trained to hit or miss a central capsule. To this end, a specific loss function is used. This network contains a reconstruction sub-network synthesizing samples of images. The reconstruction sub-network can be used as an augmentation method to avoid overfitting. HitNet uses ghost capsules to detect mislabeled data in the training set.

He et al. [12] proposed Complex-Values Dense CapsNet (Cv-CapsNet) and Complex-Valued Diverse CapsNet (Cv-CapsNet++). Both networks include a complex-valued sub-network for extracting features in different scales. Afterwards, complex-valued PCs are created out of the extracted features. Cv-CapsNet++ implements a hierarchy of three-level Cv-CapsNet model and hence produces multi-dimensional complex valued PCs.

Chen et al. [13] propose a deep capsule network combined with a U-Net preprocessing module (DCN-UN) which attempts to improve CapsNet for complex datasets such as CIFAR-10 and CIFAR-100. A convolutional capsule layer is developed based on local connections and weight-sharing strategies which allows reducing the number of parameters. DCN-UN employs a greedy strategy to develop the Mask Dynamic Routing (MDR) to improve the performance.

Ayidzoe et al. [14] introduce a less complex variant of CapsNet with an improved feature extractor. They employ a Gabor filter and customized blocks for preprocessing data leading to the extraction of the semantic information. This results in enhanced activation diagrams and learns the hierarchical information meaningfully.

Tao et al. [15] present an efficient and flexible network based on capsules referred to as Adaptive Capsule CapsNet (AC-CapsNet). This network replaces the primary capsules with an adaptive capsule layer. The adaptive values contain both the spatial information of each capsule vector and the local relationship among the neurons contained in the capsules.

This work takes a similar approach as the works explained above. However, we take into account the network accuracy and size (number of parameters) simultaneously. As we later show, we achieve the highest accuracy among the state-of-the-art networks based on capsule.

3 Background

In this Section, we review background. We explain normal and 3D CapsCells, the normal and the 3D routing algorithms, and the class-independent decoder. These units along with the CapsSum and MLCE (explained in Sect. 4) build DL-CapsNet.

3.1 Capsule Cell

Capsule Cells (CapsCells) were introduced by Rajasegaran et al. [16] as units including a combination of several Convolutional Capsule (ConvCaps) layers and a skip connection. The ConvCaps layer is a convolutional layer with its outputs reshaped to capsules. Figure 1 shows a CapsCell, which includes three ConvCaps layers. The output of the first layer is skip connected to the output of the last layer. This is done to avoid the problem of vanishing gradients. In addition, the skip-connection helps route the low-level capsules to high-level capsules. The skip-connection either includes a ConvCaps layer, or implements the 3D dynamic routing (3DR) algorithm. The former is called a normal CapsCell and the latter is called a 3DR CapsCell.

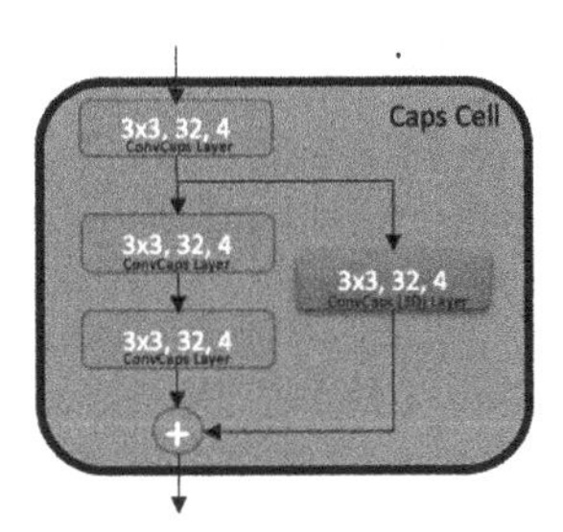

Fig. 1. The architecture of a CapsCell with $K = 3$, $D = 4$ and $N_v = 32$. This unit contains several ConvCaps layers and a skip-connection. For the 3DR CapsCells, the skip connection performs the 3D dynamic routing operation.

Each ConvCaps layer has 3 parameters: K or the kernel size, D or the number of values for each output vector (the dimensionality), and N_v or the number of vectors per spatial location of the output feature map.

3.2 Routing Capsules

In this Section we explain two routing capsule methods in subsequent layers: Dynamic Routing (DR) and three-dimensional Dynamic Routing (3DR).

Dynamic Routing. In this method, all input capsules contribute to forming any of the output capsules. DR finds a coefficient for each input capsule and works as a routing method to relate the input capsules to the output capsules. The coefficients are not trained. Instead, DR creates the output capsules iteratively during training based on the agreement between the input capsules.

Capsules in the lower-level (input capsules) need to decide how to send their vector to the output capsules (higher-level capsules). This decision is made by changing a scalar implying the weight of the capsule. This scalar is multiplied by the vector and fed as input to the output capsules. The output capsules are a weighted sum of the input capsules, with the weights determined by DR.

3D Dynamic Routing. DR routes capsules in a global manner, since all input capsules contribute to all output capsules. The 3DR algorithm, performs the routing locally. Capsules coming from nearby regions of the previous feature map are routed together to output capsules. Figure 2 depicts how 3DR groups the input capsules and routes them to the output capsules. A capsule in layer l, predicts a c^{l+1} number of capsules. Therefore, for each capsule in layer $l+1$, there are c^l predictions. s and S denote the input and output capsules respectively, and $\hat{V}$ are the intermediate votes in the routing algorithm. Like DR, the weights are iteratively inferred, and not trained.

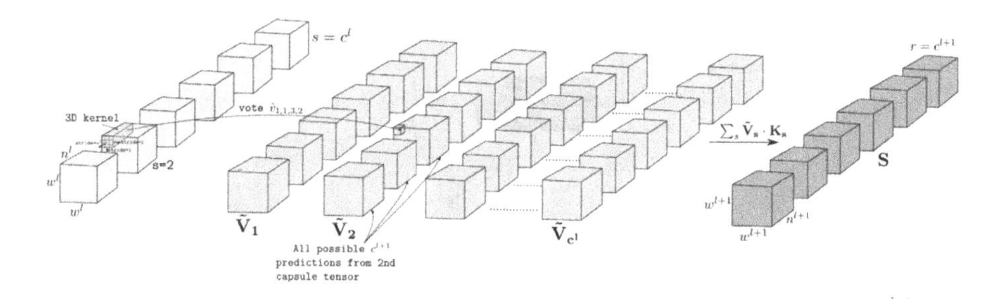

Fig. 2. 3D-Routing method. Each capsule in layer l, predicts c^{l+1} capsules. As a result, there are c^l predictions for a capsule in layer $l+1$. [6]

3.3 Class-Independent Decoder

CapsNet comes with a basic decoder based on Fully-Connected (FC) layers. The output capsules are fed to this decoder to reconstruct the input images. To regularize the training process and avoid overfitting, the reconstructed images are compared to input images. The result is considered inside the loss function (reconstruction loss). We use the class-independent decoder introduced by Rajasegaran et al. [6].

This decoder comes with two important benefits. First, it is based on deconvolution. Deconvolutional layers capture spatial relationships better than FC layers and include fewer number of parameters. Second, the decoder drops the incorrect capsules and removes them from the reconstruction process which leads to a more robust reconstruction. Sabour et al. [1] masked the incorrect capsules with zeros. The class-independent decoder, discards the incorrect capsules completely. As for all different categories (classes) there is a fixed vector of data kept and used for reconstruction, this decoder is class-independent as all classes are treated similarly. Experiments show that class-independence makes the decoder more robust [16].

4 DL-CapsNet

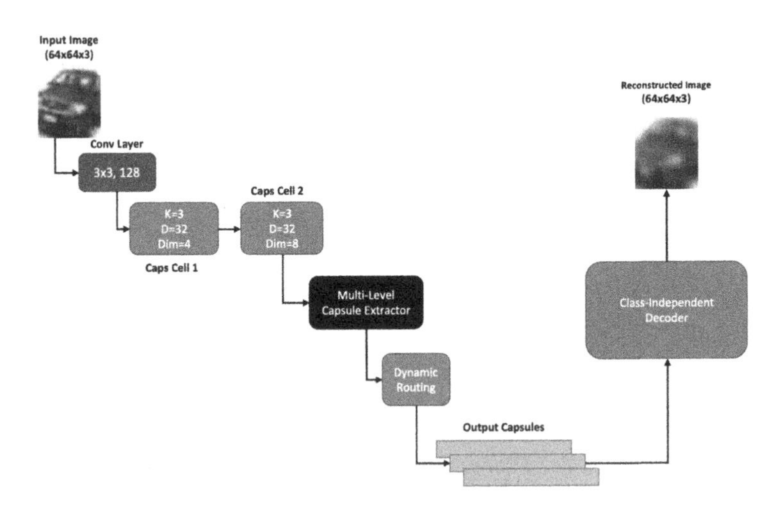

Fig. 3. DL-CapsNet architecture. The network includes two CapsCells, the MLCE module.

Figure 3 shows the architecture of DL-CapsNet. The network consists of a convolutional layer, two normal CapsCells, MLCE, the DR section and the class-independent decoder.

DL-CapsNet uses a convolutional layer to extract very low-level features. Features are reshaped to capsules. Afterwards, there are two normal CapsCells to create richer capsules. The output of the second CapsCell is fed to the MLCE module. MLCE is presented in the next Section. The capsules generated by MLCE are used to infer the output capsules of the network using DR. Similar to CapsNet, classification is done based on the output capsules. There are K capsules where K is the number of classes in the classification task, and the capsule with the highest length (L2 Norm) corresponds to the predicted class. These capsules are also fed to the decoder network.

4.1 Capsule Summarization (CapsSum) Layer

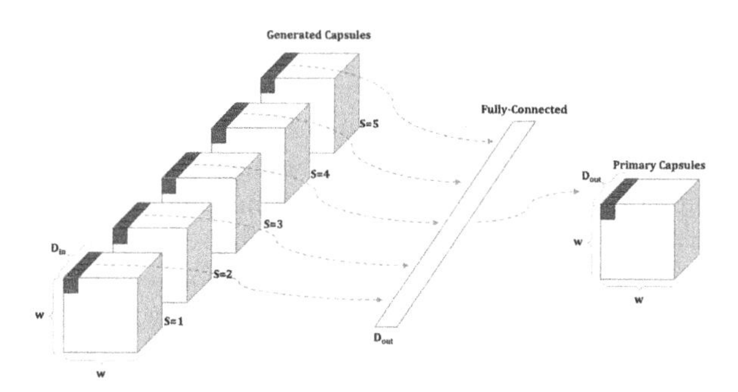

Fig. 4. The capsule summarization layer. A total of $w \times w \times 5$ generated capsules are summarized into $w \times w \times D_{out}$ primary capsules using $w \times w$ Fully-Connected (FC) layers. The first FC layer is shown.

We propose the CapsSum layer to reduce the number of generated capsules. CapsSum is inspired by the Convolutional Fully-Connected CapsNet(CFC) [10]. As stated earlier, CapsNet includes a simple feature extractor including two convolutional layers. Originally, the extracted data is directly reshaped to capsules. Alternatively, the CFC layer was proposed for translating the low-level extracted features to fewer capsules, resulting in parameter reduction and network speed-up. We customize and integrate the CFC layer in DL-CapsNet. We summarize the generated capsules by the DeepCaps network to produce a new set of PCs. To this end, we introduce a capsule summarization layer. Figure 4 shows how this layer works. The nearby generated capsules are all flattened, and fed to a fully-connected layer to produce a single capsule. This procedure is repeated for all spatial locations in the generated capsules. There are a total of the $S \times w \times w$ generated capsules each with D_{in} elements. For each spatial location (i, j) where $i, j \in [1, w]$, a total of S capsules with D_{in} dimensionality are collected, flattened and fed to a single fully-connected layer to produce a single capsule with a different dimensionality D_{out}. It is noteworthy that the figure shows the procedure for the first spatial location (1, 1). There are $w \times w$ fully-connected layers to summarize the entire generated capsules to the PCs. The proposed layer, reduces the number of capsules S times. Intuitively, each output capsule corresponds to a set of nearby capsules. We reduce all those correlated nearby capsules to a single capsule using a fully-connected layer. The reduction process includes trainable parameters (contained in the fully-connected layer). However, the reduction in the number of capsules outnumbers the increase in the parameters.

4.2 Multi-level Capsule Extractor (MLCE) Module

MLCE has two goals. First, to create a rich and robust representation of the input images using the extracted capsules, and second, reduce network size (in terms of the number of parameters).

To meet the first goal, we stack two 3DR CapsCells. 3DR is not as computationally expensive as DR due to performing the routing in a localized manner. As a result, it is possible to stack two 3DR CapsCells to make a deep representation of data.

To meet the second goal, we use CapsSum. The primary capsules are multiplied by weight matrices each of which corresponds to one of the categories in the classification task. Therefore, reducing the number of PCs results in fewer weight matrices and as a result fewer overall number of parameters. In addition, the more primary capsules are in the network, the more computationally expensive DR would become. The more primary capsules, the more time it takes for the DR algorithm to infer the output capsules from input capsules We use two instances of the CapsSum layer inside MLCE to reduce the number of generated capsules number of PCs.

MLCE is depicted in Fig. 5. MLCE infers two sets (levels) of capsules from the input capsules, and concatenates them to generate a combination of low-level and high-level capsules. This module stacks two 3DR CapsCells. First, the input capsules are fed to a 3DR CapsCell. The first set of output capsules (low-level capsules) are formed by using a CapsSum layer on top of this 3DR CapsCell. Afterwards, there is another 3DR CapsCell, and another CapsSum layer is used to summarize the high-level capsules into a few capsules (high-level capsules).

The 3DR CapsCells create a robust part-to-whole representation of data by localizing the routing process. Stacking two of these layers, results in creating a deep representation of data. In addition, combining the output of the first and second CapsCells ensures that the created capsules include both low-level and high-level information which results in an increased generalization ability of the network. On the other hand, the CapsSum layer improves the representation, reduces the number of parameters, and enhances network speed. We use a CapsSum layer after each 3DR CapsCell. The combination of 3DR routing and the CapsSum layer in two levels, enables the network to form a multi-level representation of the input image.

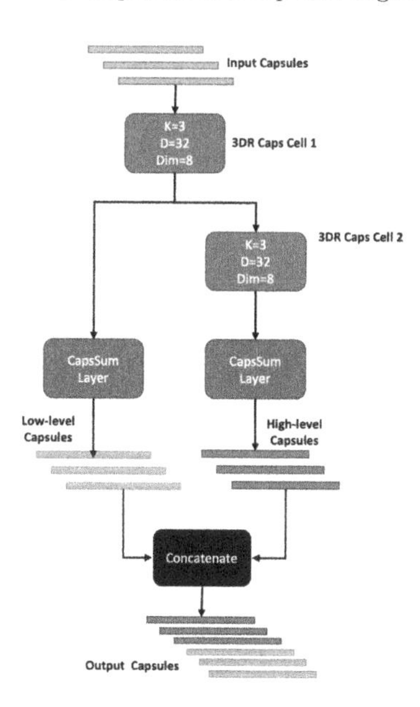

Fig. 5. The architecture of MLCE module. The module consists of two 3DR CapsCells and two CapsSum layers.

4.3 Loss Function

Our loss function is similar to the one introduced by Sabour et al. [1] (margin loss). This loss function considers penalties for incorrect predictions and disregards predictions with a very high or very low probability:

$$L_k = T_k \max(0, m^+ - ||V_k||)^2 + \lambda(1 - T_k) \max(0, ||V_k|| - m^-)^2$$

In this equation, L_k is the loss term for capsule k, T_k is 0 for incorrect class and 1 otherwise, m^+ and m^- are used to disregard high or low probabilities, and lambda is used for controlling the gradient at the start of the training.

5 Experiments and Results

In this Section, we explain the experiments and the corresponding results.

5.1 Datasets

We test DL-CapsNet for datasets commonly used for testing CapsNet and its variants: Fashion-MNIST (FMNIST), SVHN, CIFAR-10 and CIFAR-100. Testing CIFAR-100 is possible due to the small number of capsules generated by the MLCE module. For SVHN, CIFAR-100 and CIFAR-10 datasets, the input images are resized from $32 \times 32 \times 3$ to $64 \times 64 \times 3$ and for F-MNIST the original images are used throughout the experiment. We do the resizing because the images in these datasets include richer features compared to the F-MNIST dataset.

5.2 Experiment Settings

We modify and use several units introduced in DeepCaps [6]. We implement DL-CapsNet using the Keras implementation of DeepCaps[1]. We use a NVIDIA 2080Ti GPU for running the experiments. Following the DeepCaps implementation, we perform hard training in all the experiments. After training a network for the first 100 epochs, we tighten the bounds of the loss function by changing the values for m^+ and m^- and train the network for another 100 epochs. The experiments are repeated 5 times. We report the average values due to the little variation in the results. We used Adam optimizer with starting learning rate of 0.001. We also use an exponential decay ($\gamma = 0.96$) and batch size of 128.

5.3 Network Accuracy

Table 1 compares the network classification accuracy of DL-CapsNet to some state-of-the-art CNN networks (shown in the top part of the Table), and some recent and efficient variants of CapsNet (shown in the middle of the Table). DL-CapsNet and other recent CapsNet variants fall behind powerful CNN networks such as BiT-M [17]. As the Table shows, DL-CapsNet achieves competitive accuracy for all datasets compared to the state-of-the-art. To further improve the classification accuracy, we use a 7-ensemble model of DL-CapsNet. In this method, seven instances of DL-CapsNet are trained and the softmax outputs are averaged to determine the final output of the network. Using this method, DL-CapsNet reaches 91.29% and 68.36% accuracy for CIFAR-10 and CIFAR-100 datasets.

For datasets with a high number of classes such as CIFAR-100, the number of parameters can be very high. Therefore, the DR algorithm can take a long time to infer the output capsules. The 7-ensemble model of DL-CapsNet obtains 68.36% accuracy for the CIFAR-100 dataset. For the rest of the datasets i.e. CIFAR-10, SVHN and Fashion-MNIST, DL-CapsNet is among the powerful CapsNet-based networks in terms of the accuracy (DeepCaps [16] and RS-CapsNet [7]).

5.4 Number of Parameters

Table 1 shows the number of parameters besides the network inference accuracy for each dataset. Some recent and powerful CapsNet and CNN variants employ novel solutions for reducing the number of weights and include a significantly fewer number of parameters. These networks however, obtain a lower accuracy compared to DL-CapsNet. For example, DCN-UN MDR abd AC-CapsNet include 4.8M and 4.12M parameters for the CIFAR-100 dataset (compared to 11.2M in DL-CapsNet), however our proposed network achieves a slightly higher accuracy. In addition, in contrast to other works, we also report the network speed by showing how our network performs in terms of the network inference time.

[1] https://github.com/brjathu/deepcaps.

Table 1. Classification accuracy of some state-of-the art CNNs (shown on top) and the state-of-the-art CapsNet variants (shown in the middle) compared to DL-CapsNet (shown on the bottom). We obtain competitive accuracy on all datasets.

Model	CIFAR-100	CIFAR-10	SVHN	FMNIST
DenseNet [18]	82.4%/15.3M	96.40%/15.3M	98.41%/15.3M	95.40%/15.3M
RS-CNN [7]	61.14%/2.8M	90.15%/2.7M	95.56%/2.7M	93.34%/2.7M
BiT-M [17]	92.17%/928M	98.91%/928M	-	-
DA-CapsNet [8]	-	85.47%/-	94.82%/-	93.98%/-
CapsNet (7-ens) [1]	-	89.40%/11.7M	95.70%/11.7M	-
Cv-CapsNet++ [12]	-	86.70%/2.7M	-	94.40%/2.5M
CFC-CapsNet [10]	-	73.15%/5.9M	93.29%/5.9M	92.86%/5.7M
HitNet [11]	-	73.30%/8.9M	94.50%/8.9M	92.30%/8.9M
RS-CapsNet [7]	64.14%/16.8M	91.32%/5M	97.08%/5M	94.08%/5M
DCN-UN MDR [13]	60.56%/4.8M	90.42%/1.4M	-	93.33%/-
Gabor CapsNet [14]	68.17%/22.6M	85.24%/10.4M	-	94.78%/-
AC-CapsNet [15]	66.09%/4.12M	92.02%/1.26M	96.86%/1.26M	-
DeepCaps (7-ens) [16]	-	92.74%/13.4M	97.56%/13.4M	94.73%8.5M
DL-CapsNet	63.73%/11.2M	89.06%/6.8M	95.82%/6.8M	94.21%/6.4M
DL-CapsNet (7-ens)	68.36%/11.2M	91.29%/6.8M	97.09%/6.8M	94.72%/6.4M

5.5 Network Training and Inference Time

There are only few works in the CapsNet domain reporting the inference time. In addition, providing a fair comparison requires using the same base implementation for CapsNet, and the same GPU. Only the DeepCaps [6] network satisfied these two condition. Using a Geforce 2080Ti GPU, the inference in DL-CapsNet takes 2.18 ms for a single $64 \times 64 \times 3$ image of the CIFAR-10 dataset. This is 1.97x faster than DeepCaps, as it takes 4.30 ms for DeepCaps to do the same job.

6 Conclusion

We present DL-CapsNet as an efficient and effective CapsNet variant. Despite the deep structure of the network, DL-CapsNet is still a light, yet highly accurate network. Using a 7-ensemble model, DL-CapsNet achieves a competitive accuracy of 91.29% for the CIFAR-10 dataset using 6.79M parameters. With 68.36% test accuracy for CIFAR-100, DL-CapsNet performs well on complex datasets with a high number of categories.

Acknowledgment. This research has been funded in part or completely by the Computing Hardware for Emerging Intelligent Sensory Applications (COHESA) project. COHESA is financed under the National Sciences and Engineering Research Council of Canada (NSERC) Strategic Networks grant number NETGP485577-15.

References

1. Sabour, S., Frosst, N., Hinton, G.E.: Dynamic Routing Between Capsules. In: NIPS (2017)
2. Lecun, Y.: The MNIST database of handwritten digits. http://yann.lecun.com/exdb/mnist/
3. Xiao, H., Rasul, K., Vollgraf, R.: Fashion-MNIST: a novel image dataset for benchmarking machine learning algorithms, August 2017
4. Krizhevsky, A., Nair, V., Hinton, G.: CIFAR-10 and CIFAR-100 datasets (2009)
5. Xi, E., Bing, S., Jin, Y.: Capsule network performance on complex data. **10707**(Fall), 1–7 (2017)
6. Rajasegaran, J., Jayasundara, V., Jayasekara, S., Jayasekara, H., Seneviratne, S., Rodrigo, R.: DeepCaps: going deeper with capsule networks. In: Proceedings of the IEEE Computer Society Conference on Computer Vision and Pattern Recognition, pp. 10717–10725 (2019)
7. Yang, S., et al.: RS-CapsNet: an advanced capsule network. IEEE Access **8**, 85007–85018 (2020)
8. Huang, W., Zhou, F.: DA-CapsNet: dual attention mechanism capsule network. Sci. Rep. **10**, 11383 (2020)
9. Shiri, P., Sharifi, R., Baniasadi, A.: Quick-CapsNet (QCN): a fast alternative to capsule networks. In: Proceedings of IEEE/ACS International Conference on Computer Systems and Applications, AICCSA, November 2020
10. Shiri, P., Baniasadi, A.: Convolutional fully-connected capsule network (CFC-CapsNet). In: ACM International Conference Proceeding Series (2021)
11. Deli, A.: HitNet: a neural network with capsules embedded in a Hit-or-Miss layer, extended with hybrid data augmentation and ghost capsules, pp. 1–19 (2018)
12. He, J., Cheng, X., He, J., Honglei, X.: CV-CapsNet: complex-valued capsule network. IEEE Access **7**, 85492–85499 (2019)
13. Chen, J., Liu, Z.: Mask dynamic routing to combined model of deep capsule network and U-net. IEEE Trans. Neural Netw. Learn. Syst. **31**(7), 2653–2664 (2020)
14. Ayidzoe, M.A., Yu, Y., Mensah, P.K., Cai, J., Adu, K., Tang, Y.: Gabor capsule network with preprocessing blocks for the recognition of complex images. Mach. Vis. Appl. **32**(4), 91 (2021)
15. Tao, J., Zhang, X., Luo, X., Wang, Y., Song, C., Sun, Y.: Adaptive capsule network. Comput. Vis. Image Underst. **218**, 103405 (2022)
16. Rajasegaran, J., Jayasundara, V., Jayasekara, S., Jayasekara, H., Seneviratne, S., Rodrigo, R.: DeepCaps: going deeper with capsule networks (2019)
17. Kolesnikov, A., et al.: Big Transfer (BiT): general visual representation learning. In: Vedaldi, A., Bischof, H., Brox, T., Frahm, J.-M. (eds.) ECCV 2020. LNCS, vol. 12350, pp. 491–507. Springer, Cham (2020). https://doi.org/10.1007/978-3-030-58558-7_29
18. Huang, G., Liu, Z., van der Maaten, L., Weinberger, K.Q.: Densely connected convolutional networks, August 2016

Comparative Study of Scheduling a Convolutional Neural Network on Multicore MCU

Petr Dobiáš[1,2(✉)], Thomas Garbay[3], Bertrand Granado[3], Khalil Hachicha[3], and Andrea Pinna[3]

[1] ETIS, UMR 8051, CY Cergy Paris University, ENSEA, CNRS, 95000 Cergy, France

[2] ESIEE-IT, 8 rue Pierre de Coubertin, 95300 Pontoise, France
pdobias@esiee-it.fr

[3] Sorbonne Université, CNRS, LIP6, 75005 Paris, France

Abstract. Convolutional neural networks (CNNs) are progressively deployed on embedded systems, which is challenging because their computational and energy requirements need to be satisfied by devices with limited resources and power supplies. For instance, they are implemented in the Internet of Things or edge computing, i.e. in applications using low-power and low-performance microcontroller units (MCUs). Monocore MCUs are not tailored to respond to the computational and energy requirements of CNNs because of their limited resources and power supplies, but a multicore MCU could overtake these limitations. This paper experimentally compares three algorithms scheduling CNNs on embedded systems at two different levels (neuron and layer ones) and evaluates their performances in terms of the makespan and energy consumption. The results show that the algorithm called SNN outperforms other two algorithms (STD and STS) and that the scheduling at layer level significantly reduces the energy consumption.

Keywords: Convolutional Neural Network · Embedded systems · Multiprocessors · Scheduling · Simulation experiments

1 Introduction

At present, convolutional neural networks (CNNs) are mainly used for visual recognition and their utilisation is increasing. CNNs are ubiquitous and implemented in applications like image understanding, mapping, medicine or self-driving cars [15]. CNNs effectually provide solutions to different problems but they require much energy and memory space. To take advantage of their performances and deploy them in devices with limited power supplies, it is necessary to reduce their requirements. This is currently a popular research topic, many researchers focus on it [7] and it is discussed for example within the community *tinyML* (tiny Machine Learning): https://www.tinyml.org/.

K. Desnos and S. Pertuz (Eds.): DASIP 2022, LNCS 13425, pp. 69–80, 2022.
https://doi.org/10.1007/978-3-031-12748-9_6

CNNs are usually implemented on specialised hardware, such as Field-Programmable Gate Array (FPGA), or Application-Specific Integrated Circuit (ASIC), e.g. Tensor Processing Unit (TPU). They can also be used on Graphics Processing Units (GPU) but they are rather rare on Central Processing Unit (CPU). The power consumption of TPUs and GPUs is rather high and consequently these platforms are not suitable to be used on embedded systems with limited power resources. As for FPGAs, the estimate of energy consumption is not easy due to FPGA reconfigurations. The analysis of scheduling CNNs on ASICs would be feasible but it would require focusing on only one application, while we strive to obtain general conclusions. The comparison of CNN implementation on TPU, GPU and CPU was presented in [20].

Our research focuses on CPU. In particular, we consider low-power and low-performance CPUs, which are intended for embedded systems and rarely used in multicore systems. Thus, we consider a multiprocessor system based on the ARM 32-bit Cortex-M4 processors whose features make the execution of CNNs challenging. However, an achievement to implement CNNs on a battery-powered embedded system using commercial off-the-shelf processors is very promising for the use in energy-constrained applications, e.g. medical, industrial monitoring, or the Internet of Things. An example of use is wildlife image processing [6].

The aim of this paper is to evaluate three algorithms to schedule CNNs on an embedded system at two different levels (neuron and layer ones). In the paper:

- we assess the algorithm performances based on the makespan;
- we compare the results with the optimal solution provided by CPLEX solver;
- we measure the energy consumption of mathematical operations and we compute the energy consumption required by the scheduling algorithms;
- we suggest which level and algorithm are more suitable when implementing CNNs on an embedded system having limited power supplies.

The remainder of this paper is organised as follows. Section 2 summarises the related work. Section 3 presents the task and system models and the scheduling algorithms. Section 4 then introduces the experimental framework and Sect. 5 analyses the results. Section 6 concludes the paper.

2 Related Work

Firstly, this section summarises two main categories of strategies to schedule dependent tasks. Secondly, it points out techniques to reduce the computational and energy requirements of CNNs. Thirdly, it presents several energy models.

In general, scheduling of dependent tasks with communication costs is an NP-complete problem even for unlimited resources [3]. The heuristics to deal with such an NP-hard optimisation problem have already been proposed and they can be classified into several categories [13]. Two main categories are list-scheduling and cluster-scheduling [18]. In the studied context, the former method is mainly used because it is intended for a limited number of processors.

Neural networks require many resources and have high energy consumption. These constraints make their implementation on embedded systems with limited capabilities challenging. Several methods have already been proposed to reduce the number of operations, energy consumption and memory requirements. Surveys of such methods were presented in [2,7] and the authors of [11] benchmarked three embedded platforms. To show that an implementation of neural networks on MCUs is feasible, the paper [4] presents the design decisions behind TensorFlow Lite Micro (TFLM), which is a machine learning interference framework for running deep-learning models on embedded systems. The paper [19] aims to improve CNN inference throughput on heterogeneous systems.

As for the energy model, Tiwari et al. [16] presented an instruction level model. It considers that the power cost of the program can be expressed as the sum of the power costs of executed instructions and the power cost of the inter-instruction effects. These effects are due to the circuit state, resource constraints, or cache misses. The idea of counting instructions to evaluate the energy consumption is then reused, e.g. in [8]. The benefit of such a linear model is its easy computation without detailed knowledge of device characteristics. If these characteristics are available, a thorough estimation can be carried out using for example a module called *WiSeBat* [1].

All in all, performances of algorithms and neural networks were separately carried out many times but they were not analysed together. Our aim is to evaluate the makespan of studied algorithms and to assess the energy consumption required to schedule CNNs on an embedded system deployed at edge computing. Due to their limited resources, accelerators are not considered in our work.

3 Models and Algorithms

3.1 Task and System Models

Table 1 summarises notations and definitions employed in this paper.

We consider a neural network as a set of dependent tasks, which is modelled by a directed acyclic graph (DAG), as shown in Fig. 1. Each DAG is characterised by its arrival time and deadline. At *neuron level*, a DAG node represents a neuron, while at *layer level* it represents a layer (having several instances standing for neurons, independent among them within on layer), as depicted by green rectangles in Fig. 1. The arrows in the DAG represent dependencies among nodes. We implemented an algorithm generating a DAG for any neural network.

Each neuron is characterised by its execution time, expressed as the number of operations. One operation represents a time unit in our research. The number of operations depends on a layer. CNNs consist of three types of layers: convolutional one, pooling one and fully-connected one. Table 2 summarises the number of mathematical operations for one neuron for each layer.

The communication cost between two neurons is considered if two neurons are scheduled on different processors, otherwise it equals 0. Its unit is also the number of operations, although there are no mathematical operations during communication.

The system has P homogeneous processor(s) sharing the same memory. Nonetheless, the model can be easily extended to a system with heterogeneous processors, as in [21]. The preemption and rejection of nodes are not authorised.

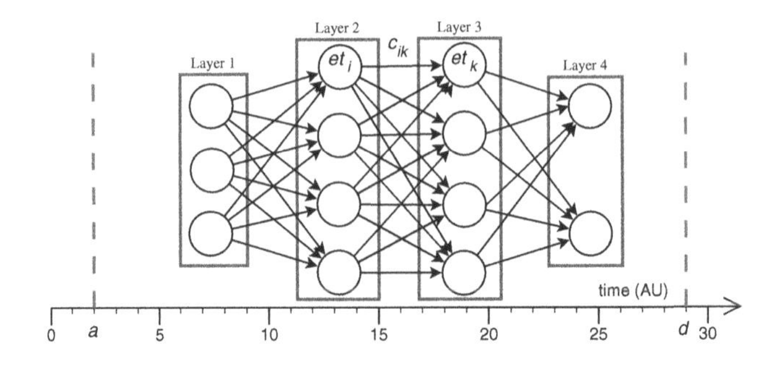

Fig. 1. Directed acyclic graph (DAG) modelling a neural network

Table 1. Notations and definitions

Notation	Definition	Notation	Definition
W	Image width	P_x	Processor x
H	Image height	a	Arrival time of the DAG
D	Image depth	d	Deadline of the DAG
F	Size of the filter	n_i	Node i
N	# of nodes	et_i	Execution time of n_i
p	# of paths in the DAG	$c_{i,j}$	Communication cost between n_i and n_j
P	# of processors	$succ(n_i)$	Set of immediate successors (children) of n_i
NL	Neuron level	$start(n_i)$	Start of the execution of n_i
LL	Layer level	$end(n_i)$	End of the execution of n_i

Table 2. Number of operations for one neuron at each layer

Layer	Number of operations for one neuron
Convolutional	$F^2 \cdot D$ multiplications, $F^2 \cdot D$ additions
Pooling	1 addition
Fully-connected	$W \cdot H \cdot D$ multiplications, $W \cdot H \cdot D$ additions

3.2 Scheduling Algorithms

In our research, we consider three algorithms based on list-scheduling.

Scheduling Node by Node. This algorithm was first introduced approximately 20 years ago [3,17] but it is still often used today, e.g. [18]. We name it *Scheduling Node by Node* (SNN) because it schedules nodes one by one in two phases for a given DAG [17]: (1) *node ordering* by their decreasing value of rank, which stands for the critical path from node n_i to the exit node n_{exit} and which is defined as follows:

$$rank(n_i) = \begin{cases} et_{exit}, & \text{if } n_i \text{ is an exit node} \\ et_i + \max\limits_{n_j \in succ(n_i)} \left(c_{i,j} + rank(n_j)\right), & \text{otherwise} \end{cases}$$

and (2) *node scheduling* to a processor minimising the execution finish time of the current node.

The complexity to compute the rank is $O\left(N^2\right)$ and the one to order nodes is $O\left(N\log(N)\right)$. To map and schedule nodes, it takes $O\left(N^2P\right)$. Consequently, the overall worst-case complexity is as reads:

$$O\left(N^2 + N\log(N) + N^2P\right) \tag{1}$$

Scheduling by Task Decomposition. The second algorithm decomposes the DAG into nodes having one or several instances and takes into account precedence constraints among nodes [5,12]. We call it *Scheduling by Task Decomposition* (STD). It also consists of two phases: (1) *node assignment of individual start times and deadlines*, which are determined proportionately to the execution time of each node with regard to the DAG arrival time and deadline; and (2) *node scheduling* to any processor respecting the start time and deadline. During the first phase, we distinguish two cases. In the former, we consider that assigned start times are fixed and, consequently, nodes are independent. In the latter case, the algorithm updates the start time of a node (child), once all of its preceding nodes (parents) are scheduled. Therefore, the nodes are not really independent because parents need to be scheduled before their children. A benefit is to reduce the makespan. To compare both versions, the former case is denoted by STD and the latter one by STDwU.

To evaluate the algorithm complexity, the assignment of individual start times and deadlines requires defining a recursive function. This recursive function considers a new path for each child and is called `ListPaths`. Its complexity (in terms of addition) is as reads:

$$\begin{cases} t(0) = 1 \\ \forall n, t(n) = \left(t(n-1) + 1\right)N \end{cases}$$

Then, the complexity to compute the duration of each path is $O\left(pN\right)$ and the one to order paths according to its duration is $O\left(p\log(p)\right)$. Next, it takes $O\left(pN\right)$ to manage start times and deadlines and $O\left(N\log(N)\right)$ to order nodes according to the "earliest deadline first" policy (and the "earliest arrival time first" policy in case of tie). Finally, the complexity of scheduling is $O\left(NP\right)$ without update or $O\left(N^2P\right)$ with update.

The overall worst-case complexity without update is:

$$O\left(N \cdot \texttt{ListPaths} + 2pN + p\log(p) + N\log(N) + NP\right) \tag{2}$$

and the one considering the update of start times is:

$$O\left(N \cdot \texttt{ListPaths} + 2pN + p\log(p) + N\log(N) + N^2P\right) \tag{3}$$

Scheduling by Task Stretching. The third algorithm is presented in [9] and is called *Scheduling by Task Stretching*[1] (STS). Its aim is to avoid parallel node

[1] The word *stretch* refers to extent all nodes of the dependent task on only one processor if possible.

execution and thus to save energy consumption by switching off idle processors. The algorithm schedules all nodes on one processor and chooses to use more processors only whenever necessary in order not to miss deadlines. Contrary to [9], we do not consider preemption in order to reduce the algorithm complexity.

This algorithm is composed of two phases: (1) *assignment of individual start times and deadlines for each node* (similar to STDwU), and (2) *node scheduling*, which distinguishes two cases. Depending on whether all nodes can be accommodated by one processor and respect their start times and deadlines, (i) nodes are scheduled to only one processor, or (ii) as many nodes as possible are placed on one processor and the remaining nodes, which were not scheduled on the first processors, are placed on additional processors.

The complexity of Case (ii) is the same as the one of STD expressed in (2). Using only one processor, the complexity of Case (i) is reduced and is as follows:

$$O\left(N \cdot \texttt{ListPaths} + 2pN + p\log(p) + N\log(N) + N\right) \tag{4}$$

3.3 Mathematical Programming Formulation

We define the mathematical programming formulation of the studied scheduling problem as follows (the notation is summed up in Table 1):

$$\text{Minimise} \max_{n_i \in \{\text{Set of nodes}\}} (end(n_i)) \text{ subject to}$$

$$\begin{cases} \textbf{1)} \text{ if } n_j \text{ depends on } n_i, end(n_i) \leqslant start(n_j) \\ \textbf{2)} \text{ if } n_j \text{ depends on } n_i, \text{ and } n_i \in P_x \text{ and } n_j \notin P_x, end(n_i) + c_{i,j} \leqslant start(n_j) \\ \textbf{3)} \ \forall \text{ time } t, \forall \text{ processor } P_x, Card(\text{nodes executing}) \leqslant 1 \\ \textbf{4)} \ (n_i \text{ and } n_j) \in P_x \Rightarrow end(n_i) \leqslant start(n_j) \text{ or } end(n_j) \leqslant start(n_i) \end{cases}$$

The objective function is to minimise the makespan, i.e. the schedule length. The first constraint guarantees that node dependencies are verified. The second constraint considers communication costs unless nodes are scheduled on the same processor. The third constraint expresses that only one node can be scheduled per processor at the same time. The last constraint forbids overlaps of nodes on one processor.

4 Experiments

4.1 Experimental Framework

In this paper, we analyse the results making use of:

- a *small CNN* consisting of 2 convolutional layers ($K = 2$ filters of size $F = 3$, stride $S = 2$ and amount of zero padding $P = 1$), 1 pooling layer (spatial extent $F = 3$, stride $S = 2$) and 1 fully-connected layer ($K = 5$ filters) and with the image size $(W \times H \times D) = (7 \times 7 \times 3)$ at input;
- *LeNet5*[2] composed of 7 layers with the image size $(W \times H \times D) = (32 \times 32 \times 1)$;

[2] http://yann.lecun.com/exdb/lenet/.

- *ShuffleNet V2* presented in [10] with the image size $(W \times H \times D) = (224 \times 224 \times 1)$
- *VGG16* presented in [14] with the image size $(W \times H \times D) = (224 \times 224 \times 3)$.

The parameters considered by our simulator are summarised in Table 3.

To compare our results, resolutions of the mathematical programming formulation (described in Sect. 3.3) were carried out in CPLEX solver[3] and computed based on the same data set.

As for the metrics, we make use of the *makespan*, which is the length of the schedule, i.e. the elapsed time between the beginning of the execution of the first scheduled node and the end of the execution of the last scheduled node. It is expressed as the number of operations as presented in Table 2. We also count the *numbers of addition, subtraction, multiplication and division* carried out during the execution of scheduling algorithm to schedule one DAG, which allows us to compute the *energy consumption* for a given algorithm.

4.2 Computation of Energy Consumption

Although more sophisticated models exist to evaluate the energy consumption, like [1], we consider a simple energy model which does not require any detailed parameters. Thus, the model is based on the number of mathematical operations. The number of instructions in assembler language required for each operation is the same for all considered operations, as indicated in Table 4. Moreover, the results of [1] show that the model we use overestimates the energy consumption. Therefore, we consider that our results represent a worst-case energy consumption and that real energy consumption is lower than this computation.

We made use of a Nucleo-144 development board with STM32L496ZG microcontroller based on the ARM Cortex-M4 processor and powered by $V = 3.3\,V$. We measured the current on the jumper JP5(IDD) using an AimTTi multimeter 1908P. For each mathematical operation, we ran the code with an infinite while loop within which a given mathematical operation (using random values) was executing. We also measured the current when no mathematical operation was implemented to compute the current dedicated to a given mathematical operation. We carried out 5 measurements and averaged the obtained values which are summed up in Table 4. The data type considered in our operations is floating-point (32-bit).

To measure the voltage, we used measurements provided by ULINKplus adapter and noted that $V = 3.305\,V$.

The energy consumption for one mathematical operation E_{op} is defined as:

$$E_{op} = V \cdot (I_{op} - I_{none}) \cdot \frac{1}{f} = \frac{3.305 \cdot (I_{op} - I_{none})}{32.000 \cdot 10^6} \qquad (5)$$

The overall energy consumption related to scheduling is the sum of all energy consumption of all mathematical operations realised by a scheduling algorithm.

[3] https://www.ibm.com/analytics/cplex-optimizer.

Table 3. Simulation parameters

Parameter	Symbol	Value(s)
Number of processors	P	1 – 10
DAG arrival time	a	0
DAG deadline[a]	d	$X \cdot$ (DAG critical path)
Execution time of n_i	et_i	see Table 2
Communication cost between n_i and n_j	$c_{i,j}$	5 operations

[a] Small CNN: $X = 20$, ShuffleNet V2: $X = 500$, LeNet5: $X = 1\ 000$, VGG16: $X = 1\ 000\ 000$

Table 4. Average current at $f = 32.000\ MHz$ and number of instructions

Operation	Measured current I_{op} (mA)	Number of instructions
None	4.286	0
Addition	4.760	6
Subtraction	4.768	6
Multiplication	4.671	6
Division	4.671	6

5 Results

In this section, we analyse the algorithm performances when scheduling different CNNs at neuron level and then at layer level. The DAG details are summed up in Table 5. The sum of execution times of all nodes (not taking into account any communication cost) is shown by dotted lines in Fig. 2a, 2b and 2d.

5.1 Neuron Level

Figure 2a depicts the makespan as a function of the number of processors for a small CNN at neuron level. When a system has only a few processors, the deadline is not satisfied, the DAG is not schedulable and no curve is plotted for a given algorithm. It can be seen that SNN achieves the shortest makespan and it is followed by STDwU. For these two algorithms, the higher the number of processors, the lower the makespan due to more resources available because the system load remains the same. Since STD does not update the assigned start times, the makespan is longer when compared to SNN and STDwU and its makespan is almost independent of the number of processors. As the makespan of STS is shorter than the overall number of operations (2038 operations), more than one processor are used. Moreover, STD, STDwU, and STS provide a solution only for systems with more than 2 processors and STD requires at least 4 processors. Actually, if a system has a lower number of processors, the deadline is not satisfied and therefore a provided solution is not considered.

Table 5. Parameters of neural networks

Parameter	Small CNN	ShuffleNet V2	LeNet5	VGG16
Critical path t_{cp} (operations)	95	56	1 560	133 435
Deadline d (operations)	1 900	280 000	1 560 000	$1.33435 \cdot 10^{11}$
Sum of exec. times (operations)	2 038	244 447	834 616	3 094 2059 008

In Fig. 2a, we also plot a magenta dotted curve showing the optimal solution based on the mathematical formulation and provided by CPLEX solver. This curve overlaps with SNN, which demonstrates that this algorithm delivers a nearly[4] optimal solution without testing all possibilities as CPLEX solver does.

The number of mathematical operations necessary to carry out by each algorithm to schedule one DAG (figure not presented in this paper) is one order of magnitude lower for SNN than for the other algorithms because SNN does not need to scour the whole DAG to determine all paths and assign individual start times and deadlines to nodes. Using these numbers of operations and (5), we compute the energy consumption. The scheduling process requires 7.593 pJ, 384.0 pJ, 384.0 pJ and 388.5 pJ respectively for SNN, STD, STDwU and STS.

While all algorithms provide results for small CNNs, only SNN do it for real CNNs. This means that STD and STS are not suitable for scheduling of real CNNs at neuron level due to their complexity expressed in (2), (3) and (4). The complexity is exponential because of the recursive function listing all paths.

5.2 Layer Level

To overcome the unsuitability of STD and STS at neuron level, we focus on scheduling at layer level, which is more coarse-grained. In general, the results at layer level are qualitatively similar to the ones at neuron level and almost independent of the size of the CNN. Only the makespan of STD can vary more significantly because it depends on the deadline value set by user.

As an example, Fig. 2b plots the makespan as a function of the number of processors for LeNet5 at layer level. We note that SNN and STDwU achieve the shortest makespan and that STD and STDwU provide a solution only for systems with more than 5 processors. The makespan of STS is always equal to the overall number of operations (834 616 operations), which means that only one processor is used. As for STD, the value of makespan is higher than the overall number of operations, which is due to the fact that the algorithm does not update assigned start times and deadlines. We remind the reader that their values were determined proportionally to the execution time of each node with regard to the DAG arrival time and deadline.

[4] Although in Fig. 2a the solution delivered by SNN overlaps with the optimal solution based on the mathematical formulation, it may happen that the optimal solution slightly outperforms SNN in different scenarios (results not presented in this paper), for example when communication costs are higher.

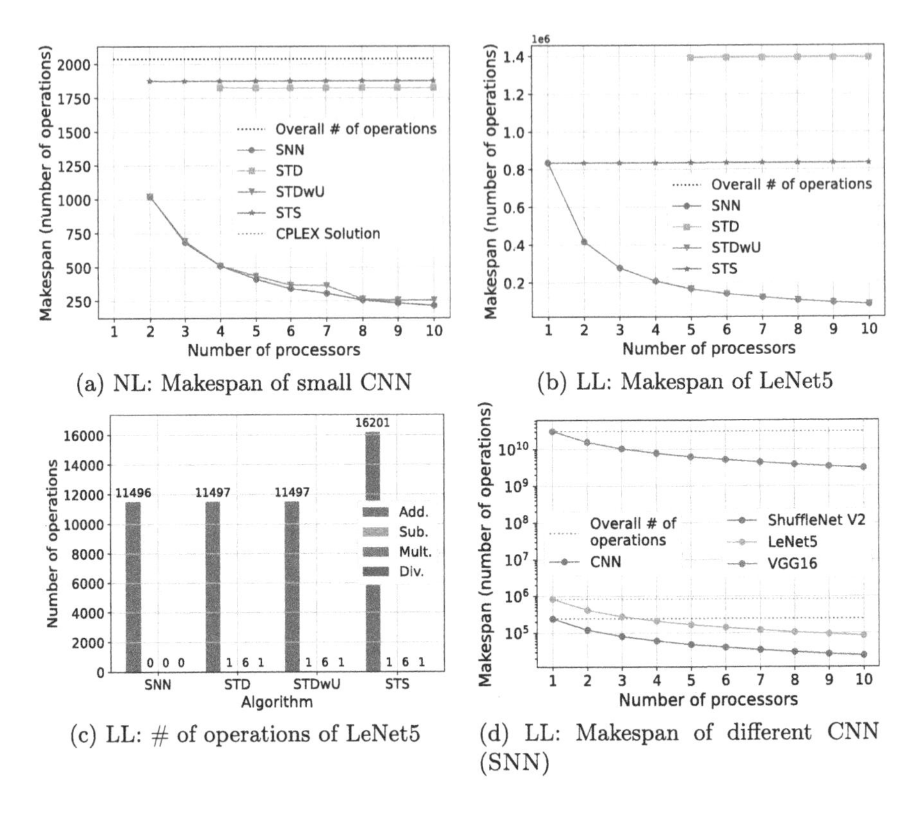

(a) NL: Makespan of small CNN

(b) LL: Makespan of LeNet5

(c) LL: # of operations of LeNet5

(d) LL: Makespan of different CNN (SNN)

Fig. 2. Comparison of different CNN at neuron and layer levels

Figure 2c plots the numbers of addition, subtraction, multiplication and division necessary to carry out by each algorithm to schedule one DAG representing LeNet5. The obtained values are independent of the number of processors. All algorithms have the same order of magnitude of the mathematical operations. The values of STS are slightly higher due to the computation of task stretching.

To show that the algorithms can deal with any CNN, Fig. 2d depicts the makespan of SNN at layer level for three different CNNs as a function of the number of processors. It can be seen that the more complex the neural network, the longer the makespan and that the trend remains the same.

5.3 Neuron Level vs. Layer Level

We thoroughly evaluated the makespan for each scenario and noted that the more complex the CNN, the smaller the difference in the makespan between neuron and layer levels. For example, the makespan of LeNet5 at layer level is longer by at most 1.0% when compared to the one at neuron level.

The role of communication cost is independent of the chosen level and depends on the algorithm. For SNN and STDwU, the higher the communication cost, the longer the makespan. While this difference is important for small CNN, it is

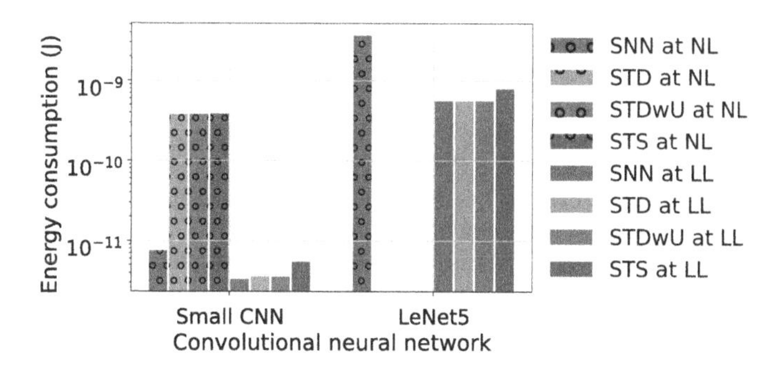

Fig. 3. Energy consumption at neuron level (NL) and layer level (LL)

almost negligible for real CNNs due to low communication to computation ratio. The communication cost when STD and STS are used has minimal impact on the makespan because these algorithms do not update the start times.

As for the energy consumption, Fig. 3 plots the required values of each algorithm for small CNN and LeNet5 at neuron and layer levels. Since the layer level is more coarse-grained, the model at layer level carries out less computations than the one at neuron layer and the energy consumption is significantly reduced: for small CNN (SNN by 56%, STDwU, STD and STS by 99%) and for LeNet5 (SNN by 85%).

6 Conclusion

The paper studied three algorithms to schedule convolutional neural networks (CNNs) on an embedded system with limited resources and power supplies. Its aim is to compare these algorithms, and to evaluate their performances in terms of the makespan and energy consumption on low-power and low-performance CPUs. These algorithms are *Scheduling Node by Node* (SNN), *Scheduling by Task Decomposition* (STD) and *Scheduling by Task Stretching* (STS).

The simulation results show that scheduling at layer level reduces the energy consumption without worsening the performances when compared to scheduling at neuron level. As for algorithm performances, SNN and STDwU function well to reduce the makespan and STS minimises the number of processors turned on. Furthermore, SNN achieves the shortest makespan and is close to the optimal solution based on the mathematical formulation and delivered by CPLEX solver. Therefore, embedded systems based on multicore central processing units (CPU) are suitable for executing CNNs.

Our future work is about to implement the studied algorithms on a platform based on RISC-V processors to further analyse their performances.

References

1. Bramas, Q., et al.: WiSeBat: accurate energy benchmarking of wireless sensor networks. In: Forum on Specification and Design Languages, pp. 1–8 (2015)

2. Capra, M., et al.: Hardware and software optimizations for accelerating deep neural networks: survey of current trends, challenges, and the road ahead. IEEE Access **8**, 225134–225180 (2020)
3. Darte, A., Robert, Y., Vivien, F.: Scheduling and Automatic Parallelization. Springer (2000). https://doi.org/10.1007/978-1-4612-1362-8
4. David, R., et al.: TensorFlow lite micro: embedded machine learning on TinyML systems (2021). https://arxiv.org/abs/2010.08678
5. Devaraj, R., Sarkar, A., Biswas, S.: Fault-tolerant preemptive aperiodic RT scheduling by supervisory control of TDES on multiprocessors. ACM Trans. Embed. Comput. Syst. **16**, 87:1–87:25 (2017)
6. Elias, A.R., et al.: Where's the bear? - automating wildlife image processing using IoT and edge cloud systems. In: IEEE/ACM Second International Conference on Internet-of-Things Design and Implementation, pp. 247–258. ACM (2017)
7. Goel, A., et al.: A survey of methods for low-power deep learning and computer vision (2020). https://arxiv.org/abs/2003.11066
8. Konstantakos, V., et al.: Energy consumption estimation in embedded systems. IEEE Trans. Instrum. Meas. **57**, 797–804 (2008)
9. Lakshmanan, K., Kato, S., Rajkumar, R.: Scheduling parallel real-time tasks on multi-core processors. In: 2010 31st IEEE Real-Time Systems Symposium, pp. 259–268 (2010)
10. Ma, N., et al.: ShuffleNet V2: practical guidelines for efficient CNN architecture design (2018). https://arxiv.org/abs/1807.11164v1
11. Qasaimeh, M., et al.: Benchmarking vision kernels and neural network inference accelerators on embedded platforms. J. Syst. Archit. **113**, 101896 (2021)
12. Saifullah, A., et al.: Parallel real-time scheduling of DAGs. Tech. Rep. WUCSE-2013-25, Department of Computer Science and Engineering, Washington University, St. Louis (2013). https://openscholarship.wustl.edu/cse_research/101
13. Shahsavari, M., et al.: Task scheduling policies in general distributed systems: a survey and possibilities. In: 22th Annual Workshop on Circuits, Systems and Signal Processing (ProRISC) (2011)
14. Simonyan, K., Zisserman, A.: Very deep convolutional networks for large-scale image recognition (2015). https://arxiv.org/abs/1409.1556
15. Stanford University: Lectures CS231n: Convolutional Neural Networks for Visual Recognition (2020). http://cs231n.stanford.edu/
16. Tiwari, V., Malik, S., Wolfe, A.: Power analysis of embedded software: a first step towards software power minimization. IEEE Trans. Very Large Scale Integr. VLSI Syst. **2**, 437–445 (1994)
17. Topcuouglu, H., Hariri, S., Wu, M.Y.: Performance-effective and low-complexity task scheduling for heterogeneous computing. IEEE Trans. Parallel Distrib. Syst. **13**, 260–274 (2002)
18. Wang, H., Sinnen, O.: List-scheduling versus cluster-scheduling. IEEE Trans. Parallel Distrib. Syst. **29**, 1736–1749 (2018)
19. Wang, S., et al.: High-throughput CNN inference on embedded ARM Big.LITTLE multicore processors. IEEE Trans. Comput. Aided Des. Integr. Circuits Syst. **39**, 2254–2267 (2020)
20. Wang, Y., Wei, G., Brooks, D.: Benchmarking TPU, GPU, and CPU platforms for deep learning. CoRR (2019). http://arxiv.org/abs/1907.10701
21. Zheng, Q., Veeravalli, B., Tham, C.K.: On the design of fault-tolerant scheduling strategies using primary-backup approach for computational grids with low replication costs. IEEE Trans. Comput. **58**, 380–393 (2009)

Design Automation and Optimization Techniques for Embedded Hardware and Software

Influence of Dataflow Graph Moldable Parameters on Optimization Criteria

Alexandre Honorat[1(✉)], Thomas Bourgoin[2], Hugo Miomandre[2], Karol Desnos[2], Daniel Menard[2], and Jean-François Nezan[2]

[1] Univ. Grenoble Alpes, INRIA, CNRS, Grenoble INP, LIG, 38000 Grenoble, France
alexandre.honorat@inria.fr

[2] Univ Rennes, INSA Rennes, CNRS, IETR - UMR 6164, 35000 Rennes, France
{thomas.bourgoin,hugo.miomandre,karol.desnos,daniel.menard, jean-francois.nezan}@insa-rennes.fr

Abstract. The integration of static parameters into Synchronous Dataflow (SDF) models enables the customization of an application functional and non-functional behaviours. However, these parameter values are generally set by the developer for a manual Design Space Exploration (DSE). Instead of a single value, moldable parameters accept a set of alternative values, representing all possible configurations of the application. The DSE is responsible for selecting the best parameter values to optimize a set of criteria such as latency, energy, or memory footprint. However, the DSE process explodes in complexity with the number of parameters and their possible values.

In this paper, we study an automated DSE algorithm exploring multiple configurations of a dataflow application. Our experiments show that: 1) Only limited sets of configurations lead to Pareto-optimal solutions in a multi-criteria optimization scenario. 2) How individual parameters impact on optimization criteria are determined accurately from a limited subset of design points. The approach was evaluated on three image processing applications having from hundreds to thousands configurations.

Keywords: Design Space Exploration · Moldable Parameter · SDF

1 Introduction

Designing signal processing applications requires an ever-increasing amount of time and resources, as well as a careful choice of the appropriate target hardware architecture, along with the corresponding software optimizations. On the hardware side, embedded systems are limited by their memory and processing power capabilities, as well as by power consumption and heat dissipation constraints. On the software side, applications for embedded systems are usually written with procedural languages such as C. As C is a relatively low-level language, it offers

This work was supported by DARK-ERA (ANR-20-CE46-0001-01).
A. Honorat, T. Bourgoin and H. Miomandre—Equal contribution.

K. Desnos and S. Pertuz (Eds.): DASIP 2022, LNCS 13425, pp. 83–95, 2022.
https://doi.org/10.1007/978-3-031-12748-9_7

the possibility to maximize the utilization of a given hardware resource through hardware-specific optimization, but at the cost of specialized cumbersome code. This widens the *software productivity gap* between the developer productivity and the code complexity required to fully exploit hardware resources [4,8].

Dataflow Models of Computation (MoCs) and associated design tools exist to bridge the *software productivity gap*. A piece of software described with a dataflow graph [11] is composed of a set of *actors*, representing computational entities, connected by First-In First-Out queues (FIFOs). A FIFO transports data tokens between *actors* which consume, process and produce said data tokens. An actor is executed when its input FIFOs contain the required number of data tokens. The production and consumption of data tokens for each actor execution is specified by a set of *firing rules*. Thus, dataflow semantics exposes task and data parallelism of data-driven computations. Design tools use the dataflow representation of an application to efficiently handle the allocation of hardware resources when deploying software on specific target hardware architectures, such as Multi-Processor System-on-Chips (MPSoCs).

This paper focuses on statically parameterized dataflow MoCs which allow both functional and non-functional behaviour of the application to be customized at compile time. We propose *moldable parameters*, that we have implemented for the first time within a dataflow MoC. Contrary to regular parameters, which hold a unique value or expression, moldable parameters are associated with a set of alternative values, each resulting in a different configuration of the application. The main contribution of this work lies in the study of how these moldable parameters impact on criteria such as latency, throughput, energy consumption and memory footprint.

In order to study the influence of moldable parameters, we have implemented a Design Space Exploration (DSE) algorithm to find the Pareto-front of multiple application configurations of the same dataflow graph. Considering a multi-objective optimization problem with independent criteria, the Pareto-front is the set of configurations providing the best trade-offs between the optimized criteria. This DSE is completed by a set of scripts used for analyzing and classifying how the different moldable parameters impact on each optimization criterion. The DSE and parameter analysis have been executed on three real-world computer vision applications: Sobel filtering, stereo-matching and Scale-Invariant Feature Transform (SIFT) [12]. Results of these analyses show that: 1) only a limited set of configurations belong to the Pareto-front, and 2) how moldable parameters impact on optimization criteria are determined accurately from a limited subset of configurations. These results lay the ground for the design of low-complexity DSE heuristics responsible for finding automatically the Pareto-efficient configurations of a set of functional and non-functional application parameters.

Related works on parameterized dataflow MoCs and resource allocation for static dataflow MoCs are presented in Sect. 2. Section 3 presents the concept of moldable parameters. Finally, Sect. 4 studies how moldable parameters impact on optimization criteria, and Sect. 5 concludes this paper.

2 Context and Related Work

In the semantics of most static dataflow MoCs, the production or consumption rate on each data port of an actor is defined by a fixed integer value [11], or sometimes by a sequence of integer values [2]. In practice, when editing a dataflow graph, it is easier to specify the production and consumption rates of actors by using symbolic expressions made of mathematical operators and functions applied to a list of pre-defined parameters associated to the graph. While the use of such parameterization mechanism is common in many dataflow frameworks, parameters are generally not part of the dataflow MoC semantics, and symbolic expressions are replaced by their resolved values before any analysis or execution of the graph. There exists a few *parametric* dataflow MoCs with a well defined parameterization semantics, as surveyed in [3]. Nevertheless, these parametric MoCs mostly integrate parameters in their semantics to support dynamic reconfiguration of the application graph during its execution. In this paper, we focus on *static* parameters whose values can be resolved statically without executing the application. After describing the model used in the present work, this section presents related work about the resource allocation for static dataflow MoCs.

2.1 Static PiSDF MoC

The parameterized dataflow MoC studied in this paper is the Parameterized and Interfaced Synchronous Dataflow (PiSDF) MoC [7] whose semantics is depicted in Fig. 1a. Configuration in the Parameterized and Interfaced Synchronous Dataflow (PiSDF) MoC is based on explicit parameters, which are nodes of the graph associated to a scalar value. The integer production and consumption rates of actors and the number of initial data token in FIFOS, known as *delays*, can be specified with expressions depending on those parameters. If all rates of an actor are evaluated to zero, it is not executed. Moreover, the Parallel and Real-time Embedded Executives Scheduling Method (PREESM) tool [13] implementing PiSDF also supports parameterized expressions of the actor execution times and energy specifications; and parameter values can be used as input argument to actor function calls in its code generation process.

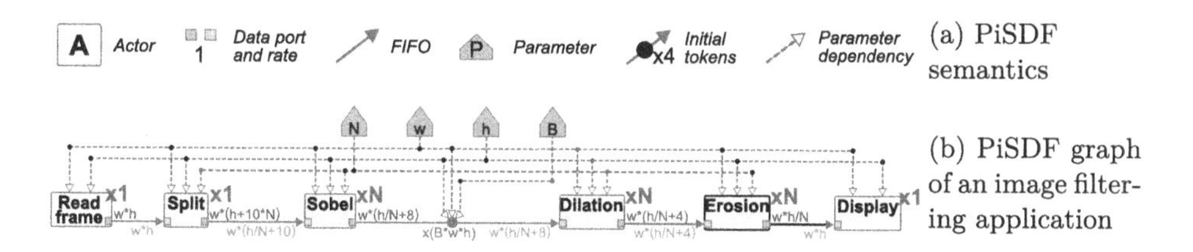

(a) PiSDF semantics

(b) PiSDF graph of an image filtering application

Fig. 1. An example of PiSDF graphical representation.

By changing the value of a parameter, it is possible to modify the functional and non-functional behaviour of an application. Non-functional parameters only impact on intrinsic properties of an application such as its degree of parallelism and granularity, tuned by data production and consumption rates; and its pipeline depth, tuned by the size of delays. Functional parameters impact on the extrinsic properties of an application, such as its Quality of Service (QoS). For example, a functional parameter can control the resolution of processed images or the bounds of an iterative process, through the modification of data production and consumption rates or number of delays. Functional parameter may also cause the execution of alternative code implementations for actors with different time or energy properties, through time and energy expressions associated to each actor and through actor parameter input. While the semantics of the PiSDF MoC also allows actors to set dynamically the values of graph parameters, in this paper we will focus only on static parameters whose alternative values are known at compile time.

Figure 1b depicts a graph implementing a simple video filtering algorithm, using elements from the PiSDF semantics in Fig. 1a. Each iteration of the graph, starting by an execution of the `ReadFrame` actor and ending by an execution of the `Display` actor, corresponds to the processing of a new frame.

Four parameters influence the behaviour of this application:

- $h, w \in \mathbb{N}^*$ are two functional parameters controlling the height and width, respectively, of the images processed by the algorithm.
- $\{N \in \mathbb{N}^* \mid h \mod N = 0\}$ is a non-functional integer parameter controlling the number of slices in which the input image is split before being processed by the N data-parallel firings of the `Sobel`, `Dilation` and `Erosion` actors.
- $B \in \{0, 1\}$ is a non-functional Boolean parameter controlling the presence of delays between the `Sobel` and `Dilation` actors. When enabled, these delays separate the computations of the graphs into two pipeline stages, thus increasing the parallelism and throughput of the application, at the expense of greater memory requirements and latency.

2.2 Resource Allocation for Static Dataflow MoCs

Software synthesis is the process translating a dataflow graph into executable code for a complex computing platform, such as a heterogeneous MPSoC [1]. To do so, software synthesis allocates all the hardware resources needed to support the execution of the dataflow graph. Among other tasks, the software synthesis is in charge of:

- *Scheduling and mapping*, which orders the individual firings of actors, and assigns these firings to the processing element handling them [5];
- *Allocating the memory* needed to store the data produced and consumed by actors near their processing elements [6];
- *Communication routing* which ensures the availability of data and the synchronization of computations [10];

- *Configuring the computing platform* appropriately to optimize the application execution, for example by tuning the Dynamic Voltage and Frequency Scaling (DVFS) configuration of the cores, or by selecting the appropriate scheduling strategy of the supporting operating system [15].

Each one of the aforementioned tasks is a complex, often NP-hard, optimization problem. Indeed, each resource allocation choice made during the software synthesis potentially impacts many optimization criteria such as the latency and throughput of the application, its energy consumption or its memory footprint. Because the resource allocation problem and optimization criteria are deeply entangled, each design decision, or each change in the dataflow graph can have intricate consequences on the different optimization criteria. For example, augmenting the parallelism of a dataflow graph by pipelining it will increase the throughput of the application, at the expanse of larger latency and memory footprint. To make all the resource allocation choices, the DSE process relies on abstract models or on means of hardware simulations for predicting rapidly the optimized criteria depending on the design decisions made. Resource allocation heuristics produce their results in a time ranging from a fraction of a second to hours, depending on their complexity and the desired quality of their outcome.

Most related work on DSE for applications modelled with static dataflow graphs assumes that the dataflow graph is fixed before entering the DSE process, thus evaluating multiple solutions given by the resource allocation solvers. When exploring different application configurations with such a DSE process, the developer must manually modify the application graph, possibly by changing its static parameters, and re-start the whole DSE process for each configuration. Only few works consider exploring design choices on the application model itself, exploiting the dataflow MoC semantics. MASES [16] is one of them; it optimizes the throughput, latency and processor utilization of applications represented with a restriction of Synchronous Dataflow (SDF) where it automatically adds software pipelining. Another tool [14] supports DSE deciding to enable actors or not, for an extended version of SDF with dynamic actors. In our work, the DSE exclusively refers to the domain of parameter configurations: it explores the multiple configurations of an application while considering a single target hardware architecture and a single resource allocation solution to each configuration. Next section introduces the moldable parameters supporting this DSE process.

3 Moldable Parameters

This section introduces *moldable* parameters, which can hold multiple expressions. Once evaluated, those expressions provide the different possible application configurations. To the best of our knowledge, this work represents the first attempt to define and integrate such moldable parameters in a dataflow MoC. After motivating the use of moldable parameters, this section presents their semantics and discusses their influence on DSE and multi-criteria optimization.

3.1 Moldable Parameters Semantics

Parameters in the PiSDF MoC may be used to set various characteristics of the application: data production and consumption rates on FIFOS, delay sizes, execution times and energy per actor firing, and even actor static integer input argument. Tuning PiSDF *regular* parameters holding a single expression is cumbersome for developers since they have to set the right expression of a parameter manually in order to run the application analysis or code generation for a specific application configuration. Instead, moldable parameters hold multiple alternative expressions so that developers do not have to set them multiple times. Most importantly, moldable parameters make it possible to automatically run analyses on all possible application configurations.

Moldable parameters are a simple extension of parameters as defined in the PiSDF semantics [7]. Each moldable parameter holds a list of symbolic expressions, separated by semi-colons. The first expression is the default one, so that moldable parameters can always be used as regular parameters. Much as regular parameters, symbolic expressions held by moldable parameters can be a mere static integer value or a complex expression depending on other parameters and using mathematical operators and functions.

In this work, we consider only *static* parameters, which means that parameter values never depend on any actor output. Moreover as parameters may depend on other parameters, cyclic dependencies are forbidden, and thus parameters and their dependencies eventually form a tree whose root parameters only hold integer values. A *parameter configuration* of the application dataflow graph is obtained by selecting and evaluating for each moldable parameter a single expression among the list of available ones. When not specified, the word *parameter* refers to both regular and moldable parameters.

3.2 Relation with Multi-criteria Optimization Problem

In order to select the most suitable parameter configuration, a developer will often consider multiple criteria: throughput, end-to-end latency, memory footprint, energy consumption, or any QoS metric. Evaluating the optimization criteria for each configuration takes from a fraction of a second to hours whereas most configurations are irrelevant, as we shall see in Sect. 4. Moreover the domain of possible configurations is the Cartesian product of the expressions of moldable parameters, so the size C of the configuration domain to explore explodes with the number of expressions held by moldable parameters. If denoting $\mathcal{P}$ the set of moldable parameters and $|p|$ the number of expressions held by $p \in \mathcal{P}$, then $C = \prod_{p \in \mathcal{P}} |p|$. Hence, there is a critical need for algorithms automating the search for the best configuration, while exploring only a subset of all possible configurations. When a moldable parameter holds only integer values, an option is to sort these values and explore only a representative sample of it.

As moldable parameters may be either functional or non-functional, their influence on the criteria to optimize are not always clear and they might compensate each other even when looking at a single criterion. In a multi-criteria optimization problem, only the points of the Pareto-front are relevant and multiple ones can be considered *optimal*. In the PREESM tool [13] supporting moldable parameters, it is also possible to automatically select a single best configuration if given a priority ordering of the aforementioned criteria. The criteria will be minimized[1], or forced to stay below a given threshold. However in the context of this paper, the criteria are neither ordered nor weighted, thus there is a priori no single best configuration. Yet the criteria to consider for the Pareto-front should be picked carefully: they should not be entirely dependent on other ones. For example energy and power are not considered together with the throughput since the power is computed by multiplying the energy by the throughput in PREESM. Next section experimentally studies the influence of moldable parameters on all criteria, except QoS ones for practical reasons.

4 Multi-criteria DSE with Moldable Parameters

This section presents experiments on three use-cases. The influence of moldable parameters on the behaviour of the chosen criteria is first evaluated through an exhaustive DSE; results are then compared with a non-exhaustive DSE.

4.1 Use-Cases: Sobel, Stereo and SIFT Applications

Three common computer vision applications have been used for experiments: Sobel, stereo and SIFT[2]. They all contain the following moldable parameters:

- `image_width`: QoS parameter holding 2 possible values for the resolution;
- `AspectRatioDenominator`: QoS parameter holding 2 possible values (but fixed for stereo);
- `parallelismLevel`: non-functional parameter holding 3 values and setting the degree of parallelism of most compute intensive actors;
- `delayRead` and `delayDisplay`: non-functional parameters enabling a pipeline stage after the image read and the result display respectively, it is used only in the corresponding delay size expressions;
- `NumeratorFrequency`: non-functional parameter simulating 12 processor frequencies and used only in expression specifying each actor timing and energy.

The execution times of actors have been measured on a JetsonTX2 board for their default configuration: maximum image size and processor frequency, no pipeline, and minimum degree of parallelism. In this work, the latency is measured as the strictly positive number of graph iterations required to process a bundle of data tokens from end-to-end, that is the full software pipeline depth controlled by the `delayRead` and `delayDisplay` parameters. The Power criterion is the sum of all actor energy specifications weighted by their number of firings, and then multiplied by the throughput. The actor timing expressions are linear

[1] For the throughput, its reciprocal is considered so that it can be minimized.

[2] Code is available upon request. For SIFT, see a similar version here: https://github.com/preesm/preesm-apps/tree/master/SIFT.

to the amount of processed data and inversely proportional to the frequency set by the `NumeratorFrequency` parameter. The actor energy expressions are linear to the amount of processed data and quadratic to the frequency.

While Sobel contains no other moldable parameter, stereo and SIFT have extra ones to enable some specific actors in the data path or to specify a QoS metric. In particular stereo can be configured with 10 different numbers of disparities used to control the accuracy of the computed depth map. The SIFT application is the most complex one containing 57 actors and 121 FIFOS dispatched into 4 levels of hierarchy and representing between 200 and 550 actor firings for each processed image, depending on the graph configuration. The moldable parameters specific to SIFT are:

- `nKeypointsMaxUser`: QoS parameter holding 9 possible values for the maximum number of keypoints to detect;
- `imgDouble`: QoS parameter enabling one resolution upsampling.

4.2 Raw DSE Results

For each configuration, the application is scheduled on an homogeneous architecture with 4 cores. A list scheduling algorithm [9] is used, followed by a static memory allocation [6]. The DSE takes a few seconds to 6 h to sequentially explore all the configurations within the PREESM framework, respectively for Sobel and SIFT. The Pareto-front is defined for 4 criteria: either Power or Energy, plus Latency, Throughput^{-1} and Memory, respectively denoted PLTS and ELTS. The Pareto-front of SIFT in the domain (Power, Throughput^{-1}, Memory) is represented in Fig. 2 for a latency value of 2, that is 2 pipeline stages.

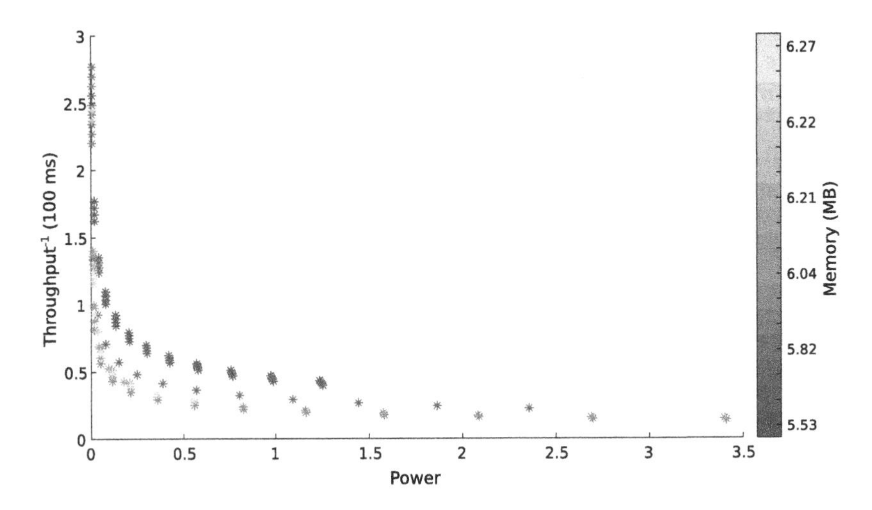

Fig. 2. Pareto-front (Power, Throughput^{-1}, Memory) of SIFT for a latency of 2.

While some purples clusters, especially around the (0.5, 0.5) coordinates, seem to be strictly dominated by blue and yellow clusters below them, they actually boast a smaller memory footprint; the Pareto-front is correct. This slightly smaller memory footprint of purple clusters comes at the cost of worse Throughput^{-1} and power consumption. Besides, the developer is most probably interested in only one point of each cluster, thus the number of relevant DSE points on the Pareto-front is even more reduced. A manual analysis of the point configurations reveals that all the clusters of points of the same colour are produced by the parameter `nKeypointsMaxUser`. Also, the parameter `NumeratorFrequency` creates the twelve clusters of each colour. Similar results were observed for other latency values and for other applications, but are not presented due to lack of space. The automatic discovery of clusters and of their relation to a specific parameter is an interesting direction for future work.

As the DSE takes up to 6 h (for SIFT), it is important to know if it really worth it to explore all the configurations. To answer this main question, we answer two corollary ones: **Q1** only the points on the Pareto-front are relevant for the developer, but how many are they among all configurations? **Q2** is it possible to explore only a subset of the configurations to get the influence of each moldable parameter on each criteria?

Q2 will be answered in Sect. 4.4 by applying the analysis detailed in Sect. 4.3 on a subset of the expressions held by moldable parameters: if expressions are mere integer, only $Npts$ are retained for the analysis. Regarding Q1, Table 1 gives the total number of configurations and the number of points on the Pareto-front for all applications and different values of $Npts$. While 20% of the points belong to the Pareto-front of the smallest application (Sobel), this percentage goes down to 1% and 2% for stereo and SIFT respectively. Note that the aforementioned percentages are for the PLTS criteria and considering $Npts = all$; they would be even lower for the ELTS criteria.

Table 1. Number of points on the ELTS and PLTS Pareto-fronts and total number of configurations for each application.

	Sobel			stereo			SIFT		
$Npts$	ELTS	PLTS	total	ELTS	PLTS	total	ELTS	PLTS	total
all	81	119	576	24	63	5760	91	244	10368
4	–	–	–	–	–	–	31	73	2048
3	19	36	144	8	29	768	23	56	864
2	11	19	64	4	25	128	12	26	384

4.3 Exhaustive Parameter Analysis

To evaluate the influence of a specific moldable parameter on a given criterion, the variation of this criterion is classified depending on the variation of the moldable parameter, with all other parameters being fixed. We classify moldable parameters in 4 categories depending on their influence on the EPLTS criteria:

- **Same** $\rightarrow$: the criterion is constant as the parameter changes.
- **Increase** $\nearrow$: the criterion strictly increases as parameter value increases.
- **Decrease** $\searrow$: the criterion strictly decreases as parameter value increases.
- **Inconsistent** ⤨: the variation of the criterion is not strictly monotonic while the variation of the parameter value is monotonic.

Each parameter-criterion pair is evaluated on all parameter configurations. If the behaviour of a criterion is consistent for all configurations, the appropriate class is assigned to the parameter-criterion pair. Otherwise, the class is **Inconsistent**. It is implemented as `Matlab` scripts which run in less than a second despite a high complexity: linear to the number of moldable parameters, to the number of criteria and to the number of configurations.

Table 2. Influence of each moldable parameter on each criterion for exhaustive DSE of SIFT. Circled results are erroneous classifications with $Npts \leqslant 3$.

Moldable parameters	Power	Latency	Throughput^{-1}	Memory	Energy
nKeypointsMaxUser	⤨	$\rightarrow$	⤨ ⊘	⤨	$\nearrow$
image_width	⤨	$\rightarrow$	$\nearrow$	$\nearrow$	$\nearrow$
parallelismLevel	$\nearrow$	$\rightarrow$	$\searrow$	⤨	⤨
AspectRatioDenominator	⤨	$\rightarrow$	$\nearrow$	$\nearrow$	$\nearrow$
delayRead	⤨	$\nearrow$	⤨	⤨	$\rightarrow$
delayDisplay	$\nearrow$	$\nearrow$	$\searrow$	⤨	$\rightarrow$
NumeratorFrequency	$\nearrow$	$\rightarrow$	$\searrow$	⤨	$\nearrow$
imgDouble	⤨	$\rightarrow$	$\nearrow$	$\nearrow$	$\nearrow$

Results of this evaluation are shown in Table 2 for SIFT. Some criteria are mostly independent of parameters, such as the Latency criterion, which only depends of the `delayRead` and `delayDisplay` moldable parameters. This particular result is expected as only these parameters impact on the pipelining of the application. At the opposite the energy evolves most of the time in the same direction as the parameters, except for the two parameters adding delays, which is also expected. On the other hand, the power and memory criteria exhibit a mostly **Inconsistent** behaviour. So power and memory are not classified the same way and while the power might be more relevant for the developer, its intricate relation with the throughput makes it harder to classify. Results on Sobel and stereo have a similar amount of **Inconsistent** parameter-criterion pairs, also linked to the power and the delays.

The parameter `imgDouble` $\{0,1\}$ controls the execution of an optional branch of the PiSDF graph. This branch upsamples the input images to find keypoints with a better accuracy. Larger input images imply more data to store and process, hence the mostly negative influence of this parameter on the criteria shown in Table 2. Consequently, as the QoS is not part of the studied criteria, the

Pareto-front does not contain any points featuring the parameter `imgDouble` with a value of 1. The same phenomenon is observed for `image_width` and `aspectRatioDenominator`: these parameters worsen the throughput and the memory and energy footprints, so there is no point having the maximum resolution on the Pareto-front. Similar results are obtained for the two other applications, with only a few exceptions about the resolution.

4.4 Faster Parameter Analysis

Here the goal is to reduce the number of configurations required to evaluate the influence of a moldable parameter on a given criterion. We observe that despite less configurations being evaluated, the results are similar to the ones described in Sect. 4.3. The same set of 4 classes defined in Sect. 4.3 is used, and the analysis is performed with the same `Matlab` scripts.

Instead of performing an exhaustive evaluation with every configuration, the number of possible expressions that a moldable parameter can be set to is capped arbitrarily according to $Npts$. Doing so, the DSE run time is greatly decreased as the domain of configurations to evaluate, being the Cartesian product of less expressions, decreases from the same ratio. This technique benefits the applications having moldable parameters holding numerous expressions. The class corresponding to each parameter-criterion pair is then evaluated on the subset of possible configurations. As all moldable parameters hold mere integer expressions in our experiments, we arbitrarily select the minimum, median and maximum values if $Npts = 3$ and equally distributed values otherwise. The choice of which expressions to keep in the general case is yet an open question.

Table 2 also shows the results of the evaluation with $Npts \leqslant 3$. For $Npts \leqslant 3$, all results are identical to the classification from exhaustive data, except for the only erroneous parameter-criterion pair displayed within a circle. The number of tested configurations for SIFT is reduced from 10368 to 864 for $Npts \leqslant 3$. For stereo, only 4 errors occur with $Npts = 2$. No error occurs for Sobel.

The classification of a parameter-criterion pair as **Inconsistent** requires an explicit divergent behaviour across multiple configurations. Consequently, limiting the DSE to a smaller sublist of moldable parameter expressions, such as $Npts = 2$, increases the likelihood of parameter-criterion pairs misclassification away from the **Inconsistent** class. Increasing $Npts$ to 4 yields a classification identical to the exhaustive analysis for SIFT, while testing only 1536 configuration out of 10368, representing 15% of the design space and an equivalent analysis speedup of almost 7x. No error occurs for stereo with $Npts = 3$, testing 432 configurations out of 5760.

5 Conclusion

This paper introduces the first use of moldable parameters in the semantics of dataflow MoCs as a way to automatically explore different configurations of an application in a multi-criteria optimization context. Results on three computer vision applications reveal that only 1 to 20% of configurations obtained with this technique belong to the Pareto-front. Finding these Pareto-efficient configurations is crucial for the developer. An analysis on how moldable parameters impact on DSE criteria shows that only a limited subset of all configuration is needed to classify accurately the influence of each parameter. This observation gives credit to the design of smart DSE heuristics capable of finding Pareto-efficient configurations without resorting to an exhaustive analysis.

References

1. Bhattacharyya, S.S., Murthy, P.K., Lee, E.A.: Software synthesis from dataflow graphs, vol. 360. Springer, NY (1996). https://doi.org/10.1007/978-1-4613-1389-2
2. Bilsen, G., Engels, M., Lauwereins, R., Peperstraete, J.: Cycle-static dataflow. IEEE Trans. Signal Process. **44**(2), 397–408 (1996)
3. Bouakaz, A., Fradet, P., Girault, A.: A survey of parametric dataflow models of computation. ACM Trans. Des. Autom. Electron. Syst. **22**(2), 1–25 (2017)
4. Castrillón, J.: Programming heterogeneous MPSoCs: tool flows to close the software productivity gap. Ph.D. thesis, RWTH Aachen University, Aachen (2013). Aachen, Techn. Hochsch., Diss., 2013
5. Castrillon, J., Leupers, R., Ascheid, G.: Maps: mapping concurrent dataflow applications to heterogeneous MPSoCs. IEEE Trans. Industr. Inf. **9**(1), 527–545 (2013)
6. Desnos, K., Pelcat, M., Nezan, J., Aridhi, S.: Pre- and post-scheduling memory allocation strategies on MPSoCs. In: Proceedings of the 2013 Electronic System Level Synthesis Conference (ESLsyn), pp. 1–6 (2013)
7. Desnos, K., Pelcat, M., Nezan, J.F., Bhattacharyya, S., Aridhi, S.: PiMM: parameterized and interfaced dataflow meta-model for MPSoCs runtime reconfiguration. In: Embedded Computer Systems: Architectures, Modeling, and Simulation (SAMOS), pp. 41–48. IEEE (2013)
8. Ecker, W., Müller, W., Dömer, R.: Hardware-dependent software, pp. 1–13. Springer, Dordrecht (2009). https://doi.org/10.1007/978-1-4020-9436-1_1
9. Honorat, A., Desnos, K., Bhattacharyya, S.S., Nezan, J.F.: Scheduling of synchronous dataflow graphs with partially periodic real-time constraints. In: Real-Time Networks and Systems. Paris, France (2020)
10. Kang, S., Yang, H., Schor, L., Bacivarov, I., Ha, S., Thiele, L.: Multi-objective mapping optimization via problem decomposition for many-core systems. In: 2012 IEEE 10th Symposium on Embedded Systems for Real-time Multimedia, pp. 28–37 (2012)
11. Lee, E.A., Messerschmitt, D.G.: Synchronous data flow. Proc. IEEE **75**(9), 1235–1245 (1987)
12. Lowe, D.G.: Distinctive image features from scale-invariant keypoints. Int. J. Comput. Vision **60**(2), 91–110 (2004)
13. Pelcat, M., Desnos, K., Heulot, J., Guy, C., Nezan, J., Aridhi, S.: Preesm: a dataflow-based rapid prototyping framework for simplifying multicore DSP programming. In: 2014 6th European Embedded Design in Education and Research Conference (EDERC), pp. 36–40 (2014)
14. Schwarzer, T., et al.: Compilation of dataflow applications for multi-cores using adaptive multi-objective optimization. ACM Trans. Des. Autom. Electron. Syst. **24**(3), 1–23 (2019)

15. Wang, J., Roop, P.S., Girault, A.: Energy and timing aware synchronous programming. In: International Conference on Embedded Software, EMSOFT 2016, p. 10. ACM, Pittsburgh (2016)
16. Yu, W., Kornerup, J., Gerstlauer, A.: MASES: mobility and slack enhanced scheduling for latency-optimized pipelined dataflow graphs. In: Proceedings of the 21st International Workshop on Software and Compilers for Embedded Systems, pp. 104–109. SCOPES 2018, ACM, NY (2018)

QoS Aware Design-Time/Run-Time Manager for FPGA-Based Embedded Systems

Alexis Duhamel[1,2](✉) and Sébastien Pillement[1]

[1] Nantes Université, CNRS, IETR UMR 6164, 44000 Nantes, France
{alexis.duhamel,sebastien.pillement}@univ-nantes.fr
[2] Capgemini Engineering, R&I Department, Rennes, France
alexis.duhamel@capgemini.com

Abstract. Due to their performance and flexibility, dynamically reconfigurable FPGA-based systems on chip find their uses in industry. Those architectures require dynamic context management of their computing resources to adapt to their environment.

Our manager dynamically changes the application quality scenarios to fulfill the system's constraints. Based on a hardware and software execution model, resources' mapping and schedule can be switched at runtime to maximize quality of service and guarantee the service execution.

In this work we intend to design such a manager with maximization of user-defined quality of service (QoS) in constrained environments and focus on continuity of service. The designed manager has been verified within a simulated environment and profiled data from an actual implementation of an H264 encoder. Results show the manager can make the targeted application run in constrained environment at the highest modeled QoS achievable without service breaks.

Keywords: FPGA · Hardware Acceleration · Reconfigurable Architectures · Runtime Management · Quality of Service · Reliability

1 Introduction

Embedded systems are used in aerospace, defense, automobile, or AI and address lots of challenges. The performance and low energy consumption of such systems are of industry's interest. Their performance, especially for parallel and data flow computing, has been shown to be significantly better than pure software execution. FPGAs have shown better performance with less energy consumption than GPUs [1], although their usage is more complex for application designers.

To cope with the resource limitation of statically reconfigured FPGAs, researchers have been using the dynamic partial reconfiguration (DPR) technique. Reconfigurable Regions (RR) can be used to reconfigure parts of an FPGA at run-time without interfering with the remaining logic elements. To try and use

K. Desnos and S. Pertuz (Eds.): DASIP 2022, LNCS 13425, pp. 96–107, 2022.
https://doi.org/10.1007/978-3-031-12748-9_8

of those hardware resources, run-time managers have been developed to optimize system's performance.

Heuristics-based managers partially solve the NP-Hard mapping and scheduling problems of managing application tasks on the system resources. They must account for resource sharing and handling of the control and communication specifications of the system's architecture. Finding near optimal solution to this problem is crucial in order to benefit from performances and flexibility offered by the DPR technique. If some works such as [2,3] have shown interest in Quality of Service (QoS) oriented management of RRs, achieving run-time continuity of service has yet to be accomplished.

In this work, we introduce a task-level hybrid design-time and run-time manager for heterogeneous hardware-software reconfigurable systems. It is able to dynamically change the application settings to maintain a high level of QoS under evolving constraint levels on the system. The proposed runtime manager considers continuity of service by trying to find a new application schedule and enforce it before the application deadline.

The obtained simulation results from an implementation of our hardware-software H264 video encoder show the constrained system is able to maintain execution of the targeted application. It manages the modeled QoS level and switches pre-evaluated mappings and schedules at run-time fast enough to keep the deadline.

The remainder of this paper is as follows. Section 2 introduces the related works. The used models are presented in Sect. 3 and 4. Our proposed approach is described in Sect. 5. Section 6 show simulation results of our experiments. Finally, conclusions are summarized in Sect. 7.

2 Related Works

The mapping and scheduling problems rely mainly on the architecture and application models. Applications are usually defined by their task graph and task properties, such as resource usage, execution times, etc. Task managers aim to find a valid mapping of the tasks on the available heterogeneous system's resources. A specific schedule for a mapping can be determined by the operating system to control the tasks and ensure their execution is deadlock-free. Task scheduling on the resources have become decisive to maximize performances [4]. Multiple works such as [5,6] have studied multi-objective mapping and scheduling problems with a focus on performance.

Because of the NP-hardness of the mapping and scheduling problem, previously cited works usually focus on a single optimized solution found at design-time. While they are capable to find near-optimal solutions, it is at the cost of execution times ranging from seconds to minutes even on a host computer. This makes the systems unable to change solution at run-time. To answer the high execution time and flexibility, hybrid managers have been introduced. Hybrid managers includes two steps: design-time and run-time. The first step executes the time-consuming heuristic on a host to generate, evaluate and prune solutions. The second step running online executes a less compute-intensive heuristic using the pruned pre-evaluated solution space. This helps reducing the timing overhead introduced by such managers. Hybrid managers and real-time approaches for self-reconfigurable systems have been studied in [7–10]. However, these approaches do not consider run-time changes in the application task graphs induced by QoS parameters nor their impact on the system.

Finally, works in [11] introduce a model for a self-reconfigurable system that considers its environment, the availability of computational resources to respect a mission plan. Its goal is to maximize QoS with evolving user's needs. This work was made with the intent to be used conjointly with the run-time manager described in [2]. The latter introduces an autonomic application version selector. At run-time, this manager checks if the execution time is within user-set bounds (or margin). If the application crosses the margin of maximum execution time threshold, the application is replaced by a less compute-intensive version. The application is upgraded if execution times are way smaller than the threshold. This work however focuses only on the quality setting aspect and doesn't answer the mapping and scheduling problems at the resource and task level.

While existing application-level approaches regarding QoS management on self-reconfigurable systems maintain near-optimal levels based on their respective objective function, the statically found solutions are designed for specific scenarios. To withstand changing environments and real world unpredictable constraints, an online management of the task/resources mappings and schedules is necessary to guarantee continuity of service of a targeted application. We address this problematic by introducing our approach based on task and resource level hybrid management with application-specific user-defined QoS goal. The emphasis is put on continuity of service at run-time through fast dynamic mapping and schedule run-time decisions.

3 Application and Quality of Service Models

In this work, the manager is restricted to one targeted application. Multi-application management would require each application to have their own QoS model, and consequently be treated separately by order of user-defined priority, and will be studied as future works. The application must be designed in such a way that a minimum QoS version of the targeted application can always be executed on the system.

The application is modeled via multiple metrics associated to its task graph. Metrics include a unique task identifier, a list of parent and children task identifiers (if any) and its profiled average execution time on software and/or hardware resource. In this model, tasks are black boxes that can be executed either on hardware and/or software resources.

The QoS model is the design-time objective function of our optimization problem, which is to maximize QoS at run-time. As our proposed manager is designed to be application-independent, this function is considered an input to provide QoS parameters for the manager to interact with. This function returns a quantifiable metric called relative preference (Q_{pref}) which reflects user perceived QoS and allows to objectively compare two application settings. Relative

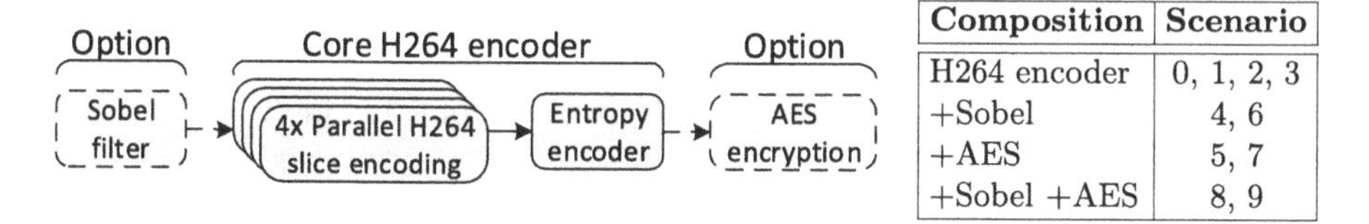

Composition	Scenario
H264 encoder	0, 1, 2, 3
+Sobel	4, 6
+AES	5, 7
+Sobel +AES	8, 9

Fig. 1. Directed acyclic graph of the used video encoder. Sobel filter and AES encryption tasks expand the graph with optional modes of executions.

preference is a dimensionless value between 0 and 1 where a score of 0 is a service break and 1 the best available setting.

We illustrate our work with a H264 video encoder example, which is representative of data flow applications. Its task graph is introduced in Fig. 1 with composition of optional modes of execution. The H264 encoder parallelly encodes 4 slices of the input image before applying an entropy encoder to the output bitstream. The encoder computes 16 × 16 pixels macroblocks with 8-bit width quantization for black and white images. The application encodes two frames at once when running at 60fps. Continuity of service can be easily represented in this case as no video frame should be dropped as a result of a failure in meeting application deadlines, similarly to soft real-time behavior.

The QoS model of the H264 application is a function of image resolution (360p or 480p), framerate (30 or 60 frames per second), optional presence of filter and/or encryption tasks in the task graph. Additionally, the task graph topology may differ between quality modes: scenarios of execution including Sobel filtering and AES encryption expand the H264 task graphs and the QoS model by introducing other parameters. A total of 10 scenarios have been defined in Table 1 based on the used parameters, with their corresponding Q_{pref} score.

The Q_{pref} score has been computed using customized Eq. 1 to simulate a user's preference, although designers are free to use a more complex equation, use empirically defined values and choose other QoS parameters to fit their application. This equation extrapolates results from [12] regarding the impact of perceived video quality and is provided as an input for the proposed method-

Table 1. Application scenarios of execution, ranked by Q_{pref} score. Each scenario is characterized by its unique combination of QoS parameters.

Scenario index	**0**	**1**	**2**	**3**	**4**	**5**	**6**	**7**	**8**	**9**
Number of tasks	5	5	5	5	6	6	6	6	7	7
Image resolution	360p	360p	480p	480p	480p	480p	480p	480p	480p	480p
Video framerate (fps)	30	60	30	60	30	30	60	60	30	60
Sobel filter	No	No	No	No	Yes	No	Yes	No	Yes	Yes
AES encryption	No	No	No	No	No	Yes	No	Yes	Yes	Yes
Q_{pref} **score**	0.18	0.25	0.35	0.53	0.59	0.59	0.76	0.76	0.82	1.00

ology. With '360p' and '480p' (similarly with 'No' and 'Yes') corresponding to indices 0 and 1 respectively, we use parameters from Table 1. With r as image resolution, f as video framerate, s and a as presence of sobel filtering and AES encryption respectively:

$$Q_{pref}(r, f, s, a) = \frac{0,75.r + 2,5.f.10^{-2} + s + a}{4,25} \tag{1}$$

Finally, we refer to continuity of service as a system's ability to deliver a minimum level of service (represented by its QoS model) of a targeted application at run-time. The run-time manager focuses on a single application (represented as a task graph) and shares the hardware and software resources with other context applications that aren't managed. As the latter occupy resources, their effect on the system is seen by the run-time manager as a constraint on the resources usage. This may cause an increase in execution times which could lead to non-respect of execution time deadlines. To ensure continuity of service, the manager dynamically changes the targeted application's QoS scenario to find a pre-evaluated solution that's able to compute the application while respecting the user-defined constraints, which in this paper is the execution time deadline of the application.

4 System Model

We consider a generic targeted system architecture introduced in Fig. 2, comparable to works such as [13]. In this work, we target a Zynq-7000 PYNQ-Z2 board containing a FPGA, a CPU, and on-chip memories.

The FPGA part contains RRs and their respective control and communication static interfaces. Those interfaces are responsible for local management of low-level control operations and data transfer between the RR logic and the hardware communication infrastructure. The latter consists of a switch transferring data between RRs and software threads via direct memory access. Management of dynamic reconfiguration is done via the hardware manager which embeds an ICAP controller such as [14].

Point-to-point communication between RRs is considered to reduce communication delays between hardware implementations of tasks. Such communications channels are controlled and monitored via a control register in the hardware

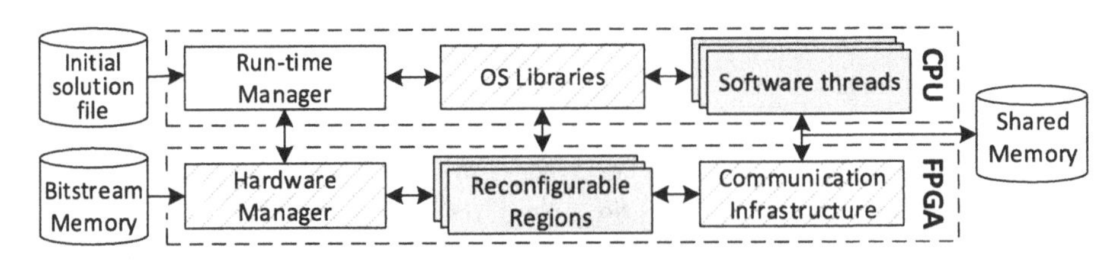

Fig. 2. Overview of the targeted reconfigurable SoC architecture. Colored blocks execute application-related functionalities. Striped blocks are out of this paper's scope. (Color figure online)

manager. This ensures the OS libraries are aware of this daisy-chaining. When receiving a new configuration order, the hardware manager configures the communication infrastructure appropriately. Usual Linux OS services are used when dealing with software application threads.

On the CPU, OS libraries such as Linux-based FOS [13] that we use in this work controls the RRs using their respective control and status register. Dedicated application software threads allow software task implementations execution. Finally, the run-time manager is our proposed software decision making greedy-based algorithm. It sends its reconfiguration requests to the hardware manager after making decision.

We consider heterogeneous RRs: they are different in available logic elements and some tasks cannot be instantiated in all RRs. Because we use a design-time mapping and scheduling algorithm to pre-evaluate solutions, the added complexity of using heterogeneous RRs does not impact the run-time algorithm. We target coarse-grained tasks being instantiated on the system. Coarse-grained tasks is the most popular granularity and represent functions of an application. This choice is motivated by a reduction of combinatorial explosion brought by the HW/SW partitioning problem, as coarse-grain reduce the number of nodes in a task graph. Also, coarse-grain tasks impose a lesser logic element overhead versus fine-grained tasks models [15].

Similarly to the application model, we define metrics for the system resources. For each computational resource is given the following metrics: nature of the computational resource (hardware or software), a unique resource identifier, average external memory access delays and reconfiguration time, including control and interface timing overhead. Additionally, average communication delays between computational elements are considers when computing schedules.

5 Proposed Hybrid Manager

Hybrid run-time managers have been introduced to cope with the NP-hardness of the dynamically reconfigurable SoCs' management. In this section, we describe our proposed hybrid runtime manager, whose goals are i) to maximize the QoS based on the user-designed QoS model at run-time without breaking the service, and ii) to quickly find a suitable pre-evaluated solution to enforce it.

The manager comprises a design time and a run-time steps. The design time step is responsible for mappings and schedules (i.e.; solutions) generation and solution space reduction, after which an initial solution file is created. The run-time step is responsible for the actual management of the system according to the current monitored execution context. It is responsible for optimizing the QoS with the pruned solution set from the initial solution file and information given by the application designer. Figure 3 presents an overview of our approach.

5.1 Design Time Solution Generation and Evaluation

Given the system, application and QoS models, the mapping and scheduling problems can be solved with heuristics to generate solutions by assigning tasks

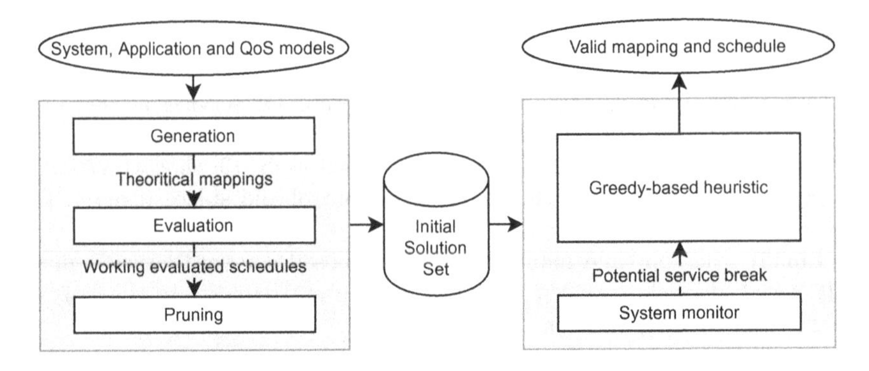

Fig. 3. Proposed hybrid manager overview. The design time and run-time step are executed respectively on a host computer (blue) and on the embedded system (orange). (Color figure online)

to the resources and computing a schedule. Input QoS and applications models filter obviously invalid solutions, such as a mapping with a task located in a RR that has no hardware implementation. We encourage the use of heuristics such as genetic algorithms for solution generation as the exhaustive method generates a large number of solutions growing exponentially with number of tasks and RRs that will most likely be pruned off during the solution space reduction step anyway.

The generated solutions are then evaluated by finding a schedule for the generated mapping. The schedule duration is obtained considering the tasks' profiled execution time on their assigned resource, the assigned resource reconfiguration time (if any) and communication channels average delays. The scheduler makes use of the reuse and pre-fetch techniques as introduced in [16]. The reuse technique consists in not reconfiguring an already implemented task if it is the only scheduled task for a resource. The pre-fetch technique consists in reconfiguring an idling resource before the task is called. This parallelizes the reconfiguration process with the execution of tasks and reduces the timing overhead introduced by reconfiguration times. We also consider FPGA constraints such as communication and reconfiguration delays in the scheduling process. After evaluation, a solution is characterized by:

- Its corresponding QoS scenario and Q_{pref} score from the QoS models;
- Its pre-evaluated profiled execution time;
- A mapping matrix M of dimension $n \times m$, with m being the number of resources and n the number of tasks. Element $M_{i,j}$ is 1 if task i is assigned to resource j, else 0;
- A communication channel matrix C of dimension $m \times m$ whose element $C_{i,j}$ is 1 if a task assigned to resource i requires a communication channel with a task assigned to resource j, else 0.

Finally, a pruning of the solution space ends the design time step. Solutions that don't satisfy execution time deadlines are removed. For each QoS scenario, the solutions are ranked by their execution time, and the lowest ranked are discarded in such a way that only a user-defined number of solution remains for each setting. This ensures remaining solutions only include viable best solutions for a given scenario.

5.2 Run-Time Management

While the initial solution set was evaluated at design time on free resources, their actual run-time implementation will have to share the resources with context applications. This affects the system in a way that could make those solutions non compatible with the current context. This leaves the run-time problem to find which best viable solution within the set can be implemented in the current context.

The SoC's CPU handles the run-time management. It is enabled whenever the system monitor detects that the influence of other running programs on the system may cause a service break in the next iteration. The system monitor is a representation of the OS library that checks the execution status of tasks. It keeps track of system resources in order to extract current CPU and FPGA implementation application metrics. Finally, it looks for constraints violations. The system monitor can be customized to the user's requirements, such as monitoring energy usage, chip temperature, and so on. In this work, the system monitor checks the execution time metric for each encoded frame. If the system monitor observes execution times crossing the deadline minus a user-defined margin (similarly to [2]), a downgrade is requested to the heuristic to find a new solution. If no such request is sent the run-time manager defaults to check for an application upgrade.

Application downgrade refers to the process of lowering the QoS in order to avoid service breaks. Assuming that no implementation of the present setting would allow the program to meet deadlines, a less resource-intensive setting is chosen. An application upgrade, on the other hand, tries to increase QoS. Downgrades must be made as quickly as possible in order to execute the newly found solution on the current iteration. Upgrades can be delayed until the following iteration as the current application implementation runs fine but sub-optimally. This allows more time to identify a better solution and avoid making frequent QoS-impacting changes. Thus high upgrade decision times do not cause service breaks.

Once the decision to upgrade or downgrade has been made, the run-time manager uses a Greedy-based heuristic to identify a new compatible solution of the application. The initial solution file's reduced solution space contains a database of solutions meant to execute the managed application in the targeted QoS scenarios and using different resource combinations. At equal QoS score, some solutions provide a shorter application makespan or require more or less resources than others.

The greedy-based heuristic parses the pruned solution set with an upgrade or downgrade goal. Some solutions may no longer be implementable because their required resources are being used by other context applications, and including the targeted application tasks in the schedule of this resource would cause the application to miss deadlines. We refer to remaining solutions that can be instantiated as compatible solutions. Because they favor exploitation over exploration, greedy algorithms typically find solutions faster than optimal methods. The algorithm stops when it has found one viable solution. As future works, a second round of search using a more compute-intensive algorithm could be done to look for a more long-term compatible solution.

For each solution, the greedy-based heuristic checks if a solution fits given the current system resources occupied by context applications and services. To do so, it tests the pre-evaluated schedule of the solution on the current monitored state of the system's resources. If the whole schedule, which includes the solution and the context application schedule, fits within the execution time deadline, it is judged a compatible solution and the heuristic stops.

6 Experimentation and Results

6.1 Environment Model

The simulated system's architecture comprises of 4 heterogeneous RRs. The targeted board is a PYNQ-Z2 board which embeds Xilinx Zynq-7000 FPGA and 650 MHz dual core ARM A9 CPU SoC. Partial bitstreams size were 650 KB on average. Using 380 MBps reconfiguration bandwidth from [14], we profile reconfiguration times at 1.70 ms on average. We measured an intra-software and intra-hardware communication delay between resources equal to 0.01 ms, and 0.26 ms whenever a transfer between hardware and software occurs using Xilinx's DMA IP and the board's DRAM.

We model the effects of context applications constraints on the system to verify the feasibility of our approach. The context execution time for each resource represents the amount of time we consider it to be completely unavailable for the application. Random context execution simulates the unpredictability for the runtime manager of those context constraints. The context execution time of each resource is simulated using a normal distribution centered at half the deadline d and scaled for values between $d/4$ and $3d/4$.

6.2 Benchmark Application

We used the H264 encoder as benchmark application as defined previously. H264 tasks have been profiled in both hardware and software domain whenever possible. The encoder's performance has been measured for two resolutions (480p and 360p) with a deadline of 27 ms (margin considered). Its hardware performance while running with a 100 MHz clock has been measured on the target board using Xilinx's Integrated Logic Analyzer (ILA) on a static FPGA design. Execution times of the software implementation of the encoder's tasks were measured on the targeted CPU using the C time library. The FPGA implementation

Table 2. Number of solution in the initial solution file, ordered by scenario.

Scenario	0	1	2	3	4	5	6	7	8	9
Nb of solution	7871	5223	6947	2480	6366	3564	1215	2016	2215	110
Min. exec. (ms)	5.16	10.27	7.84	15.52	9.13	13.68	21.36	16.81	16.67	25.57
Avg. exec. (ms)	7.48	13.86	9.64	19.09	10.50	17.24	23.35	19.01	21.50	26.30

outperforming the CPU due to their specifications, we cannot use a full software implementation. However, some scenario considering partial software implementation, albeit sub-optimal, can be executed to alleviate FPGA constraints in some cases.

An exhaustive solution generation has been executed to generate all possible solutions at design time to avoid local optimums introduced by heuristics. Because of combinatorial explosion, exhaustive approaches scale really badly with the number of tasks in the task graph, number of resources, and number of QoS scenarios. As mentioned in Subsect. 5.1, we encourage the use of heuristics instead. Given our application and system models, 1'156'242 solutions have been generated and pre-evaluated in 110 s on a host computer. After pruning, 38'007 solutions were kept (96.71% reduction) in the initial solution file. Table 2 shows the number of solution for each scenario, along with minimum and average execution times.

The simulation environment and algorithms were written in Python for prototyping purpose and executed on the PYNQ-Z2's CPU.

6.3 Results

Table from Fig. 4 shows average resulting solution relative preference and decision time from our run-time manager over 100'000 encoder iterations compared to the exhaustive optimal on the pruned set of solutions. As expected, the exhaus-

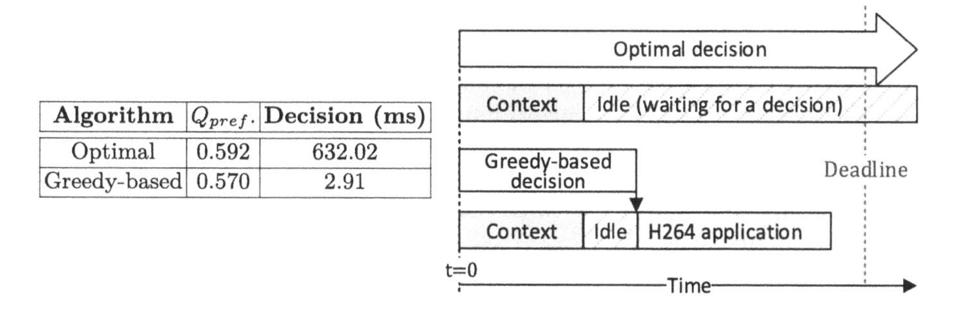

Algorithm	Q_{pref}.	Decision (ms)
Optimal	0.592	632.02
Greedy-based	0.570	2.91

Fig. 4. On the left: average relative preference Q_{pref} and decision time by algorithms running on the targeted board's CPU, averaged over 100'000 iterations. Illustrated on the right: impact of the decision time on the lateness of the H264 application execution.

tive algorithm returns a much higher decision time and fails to find solutions fast enough to avoid service breaks (i.e.; the application cannot run after decision and still meet the deadline, see illustration in Fig. 4). Our Greedy-based algorithm manages to return an average relative preference 3.71% below optimal with a much shorter decision time.

When performing a downgrade, we add the decision time to the makespan of the found solution as the application starts running immediately after the solution is found. Fast downgrading is required to allow continuity of service, as a decision too long causes the application to break the deadline. Application upgrades aren't critical in this regard, as the application already runs fine, although at a lower QoS scenario than optimal, and the found solution can be executed at the following iteration. With this in consideration, no service break were recorded on 100'000 iterations.

The run-time manager is able to run the targeted H624 encoder despite the context tasks sharing the encoder's resources. Solution changes frequently happened as simulated constraints changed at every iteration, which is an extreme scenario as FPGA implementations tend to have fixed execution times. Such changes in the mapping and scheduling can happen without seen effects on the relative preference: the user cannot perceive when the H264 encoder runs on all four RRs, or a mix of RRs and CPU. Differences between optimal and greedy-based highlighted in this table focuses on the perceived QoS modeled by the Q_{pref} value.

Compared to other approaches such as [3,8] that aims to maximize user-defined QoS, ours focuses on a task and resource level mappings and schedules, taking in consideration the decision time to make a fast decision and ensure continuity of service.

7 Conclusion

Our run-time manager aims to guarantee the execution of an application running on a constrained systems. The manager offers task and resource-level run-time management using relevant application and system models. The system architecture's specificities are considered at design-time during solution generation. We make use of a greedy-based heuristic to make fast decisions and ensure the found solution can be executed on the remaining time before deadline.

Results show the proposed algorithm is capable of avoiding service breaks while holding near-optimal QoS, illustrated by user-defined QoS scenarios, and to adapt to the context induced constraints. Our approach is capable of dynamically managing hardware and software schedule implementations of the targeted application at run-time.

Future works include an implementation of the targeted architecture to study the computational overhead of the run-time managers integrated on the FOS FPGA designed OS and architecture. Multi-application management will also be studied, as well as a way to preemptively find better schedules than greedy-based to avoid frequent QoS-impacting schedule changes.

References

1. Nurvitadhi, E., et al.: Can FPGAs beat GPUs in accelerating next-generation deep neural networks? In: FPGA (2017)
2. Gueye, S.M.K., Rutten, E., Diguet, J.P.: Autonomic management of missions and reconfigurations in FPGA-based embedded system. In: Conference on Adaptive Hardware and Systems (2017)
3. Roy, S.K., Devaraj, R., Sarkar, A., Senapati, D.: SLAQA: quality-level aware scheduling of task graphs on heterogeneous distributed systems. ACM Trans. Embed. Comput. Syst. **20**, 1–31 (2021)
4. Iguider, A., Bousselam, K., Elissati, O., Chami, M., En-Nouaary, A.: Heuristic algorithms for multi-criteria hardware/software partitioning in embedded systems codesign. Comput. Electric. Eng. **84**, 106610 (2020)
5. Sun, Z., Zhang, H., Zhang, Z.: Resource-aware task scheduling and placement in multi-FPGA system. IEEE Access **7**, 163851–163863 (2019)
6. Biondi, A., Buttazzo, G.: Timing-aware FPGA partitioning for real-time applications under DPR. In: Adaptive Hardware and Systems (2017)
7. Chillet, D., Eiche, A., Pillement, S., Sentieys, O.: Real-time scheduling on heterogeneous system-on-chip architectures using an optimised artificial neural network. J. Syst. Architect. **57**(4), 340–353 (2011)
8. Weichslgartner, A., Wildermann, S., Götzfried, J., Freiling, F., Glaß, M., Teich, J.: Design-time/run-time mapping of security-critical applications in heterogeneous MPSoCs. In: SCOPES (2016)
9. Spieck, J., Wildermann, S., Teich, J.: Scenario-based soft real-time hybrid application mapping for MPSoCs. In: Design Automation Conference (2020)
10. Abdi, A., Zarandi, H.R.: HYSTERY: a hybrid scheduling and mapping approach to optimize temperature, energy consumption and lifetime reliability of heterogeneous multiprocessor systems. J. Supercomput. **74**(5), 2213–2238 (2018)
11. Hireche, C., Dezan, C., Mocanu, S., Heller, D., Diguet, J.P.: Context/resource-aware mission planning based on BNs and concurrent MDPs for autonomous UAVs. MDPI Sens. **18**, 4266 (2018)
12. Debattista, K., Bugeja, K., Spina, S., Bashford, T., Hulusic, V.: Frame rate vs resolution: subjective evaluation of spatiotemporal perceived QoS under varying computational budgets. Comput. Graph. Forum **37**, 363–374 (2017)
13. Vaishnav, A., Pham, K.D., Powell, J., Koch, D.: FOS: a modular FPGA operating system for dynamic workloads. ACM Trans. Reconfigurable Technol. Syst. **13**, 1–28 (2020)
14. Sultana, B., et al.: VR-ZYCAP: a versatile resourse-level ICAP controller for ZYNQ SOC. Electronics **8**, 899 (2021)
15. Jain, A.K., Maskell, D., Fahmy, S.: Are coarse-grained overlays ready for general purpose application acceleration on FPGAs? In: IEEE International Conference on Dependable, Autonomic and Secure Computing (2016)
16. Ramezani, R.: A prefetch-aware scheduling for FPGA-based multi-task graph systems. J. Supercomput. **76**(9), 7140–7160 (2020)

Fixed-Point Code Synthesis Based on Constraint Generation

Sofiane Bessaï[1(✉)], Dorra Ben Khalifa[1], Hanane Benmaghnia[1], and Matthieu Martel[1,2]

[1] University of Perpignan, LAMPS Laboratory, 52 Av. P. Alduy, Perpignan, France
{dorra.ben-khalifa, Hanane.Benmaghnia,matthieu.martel}@univ-perp.fr, sofiane.bessai@etudiant.univ-perp.fr

[2] Numalis, 265 Av. des États du Languedoc, Montpellier, France

Abstract. Fixed-point arithmetic is a well-known alternative to floating-point arithmetic on embedded systems. It is used to reduce some computation costs in terms of speed and power consumption on certain platforms, e.g. medical devices, cars, and robots. In this article, we present POPiX, a novel fixed-point program synthesis tool based on static analysis. The originality of our method is to solve a system of constraints generated from the program source code. Thus, the solution of our constraints gives the new fixed-point formats while accomplishing the accuracy required by the user. Basically, POPiX takes as input an imperative program running in floating-point arithmetic and synthesizes a new program coupled to a fixed-point library relying on integers only. We evaluate POPiX on a collection of floating-point benchmarks coming from FPBench. Results demonstrate the efficiency of our analysis by achieving memory savings up to 75% with energy savings up to 3.5×.

Keywords: Fixed-point arithmetic · code synthesis · precision tuning · linear programming · static analysis

1 Introduction

Floating-point arithmetic is the dominant approximation to represent a large spectrum of real numbers. Although it offers better precision, programmers do not always need the high level of accuracy offered by the largest floating-point formats. In addition, owing to its complex internal circuitry and the increased memory requirements, floating-point arithmetic can be exorbitant in terms of speed and power consumption on certain platforms such as mobile phones, video game consoles, and digital controllers. To bridge this gap and since many embedded architectures can be implemented using very low bit-width numbers, the solution is to deploy the fixed-point arithmetic as an alternative to the floating-point one as it can be efficiently realized using integer arithmetic. Fixed-point

This work is supported by La Région Occitanie under Grant GRAINE - SYFI: https://www.laregion.fr.

K. Desnos and S. Pertuz (Eds.): DASIP 2022, LNCS 13425, pp. 108–120, 2022.
https://doi.org/10.1007/978-3-031-12748-9_9

numbers in a certain format maintain a fixed divisor (so the name fixed-point). With this kind of arithmetic, the number of bits splits into two parts, respectively named integer part and fractional part with a radix point that lies between them. Besides, many fields have revived their interest in fixed-point arithmetic when searching for cost effective hardware processors with less design effort. For instance, machine learning algorithms and models have been recently implemented using fixed-points with little accuracy loss [8, 10]. The conversion of code designed for embedded systems from floating to a fixed-point equivalent version is a long-established problem addressed in the litterature [3–6, 12]. Nevertheless, rarely we found tools (except for [4]) that deal with adjusting the floating-point formats in the original program before the conversion pass. Practically, programmers do not always need the high level of accuracy in the floating-point formats. However, manually adapting the precision of the variables may require considerable programming skill and application domain expertise. Given this consideration, automating the task of adjusting the program variables precision to improve its performance characteristics, before conversion, can help programmers to achieve performance benefits. Generally, this process of adjustment, also called precision tuning, involves a user requirement of accuracy and a semantic analysis of the program. The benefit of this tuning phase is to provide an optimized mixed-precision programs and pieces of information indicating the most suitable fixed-point data formats in the converted program.

The goal of this article is to propose POPiX: a new tool to transform a given numerical floating-point program into semantically equivalent one that exploits fixed-point computations with integers only. The key idea of this work relies on a semantic modelling of the numerical errors propagation throughout the floating-point program. In order to achieve the conversion, POPiX combines two fundamental steps. The first step consists in generating an Integer Linear Problem (ILP) from the original program in order to obtain the minimal formats (number of bits before and after the radix point) which fulfill the accuracy requirements. Basically, this is done by reasoning on the most significant bit and the number of significant bits of the values. The ILP problem can be optimally solved in one shot by a classical linear programming solver (LP) with no iteration. To the best of our knowledge, this is the first work that interests in statically synthetizing fixed-point code using an ILP formulation of the program. At the end of the tuning phase, the second step collects each information provided by the former step. The tool internally calls a fixed-point library to convert the associated indications into ones that exploit fixed-point computation with the number of bits required for each of the integer and the fractional parts. We evaluate POPiX on a set of benchmarks coming from the FPBench community. The POPiX source code, and all the data and results presented in this article are publicly available at: https://github.com/sbessai/popix.

The remainder of this article is organized as follows. In Sect. 2 we present a motivating example describing our approach. Section 3 provides all the technical details about our new developed fixed-point code synthesis framework. Experi-

```
1    [...]
2  NumTuples|10| = 40.0|5,10|;
3  create_vector(x,40);
4  create_vector(y,40);
5  create_vector(z,40);
6    [...]
7  while(i<NumTuples) {
8   x2|-6,10|=x[i]|-3,10| * x[i]|-3,10|;
9   y2|-1,11|=y[i]|-1,11| * y[i]|-1,11|;
10  z2|-6,10|=z[i]|-3,10| * z[i]|-3,10|;
11  tm|-1,10| = x2|-6,6| + y2|-1,11|
12  + z2|-6,5|;
13  magSqrt[i]|-1,10|=sqrt(tm)|-1,10|;
14     i|5,10| = i|5,11| + 1.0|0,6|;} ;
15   [...]
16 while(n<NumTuples) {
17  res|0,11| = 0.0|0,11|;
18  i|0,10| = 0.0|0,10|;
19  while (i<LpfFiltLen) {
20     if (n-i>=0.0) {
21        aux0|2,10| = lpfCoeffs[i]|2,10| *
22        magSqrt[n-i]|-1,10|;
23        res|8,10| = res|8,11| + aux0|2,9|;};
24     i|3,9| = i|3,10| + 1.0|0,7|;} ;
25     n|5,8| = n|5,9| + 1.0|0,4|;};
26     [...]
27 require_nsb(res, 8);
```

```
1     [...]
2  int NumTuples; // <0,10>
3  int16_t aux0;  // <2,10>
4  int16_t res; // <8,11>
5  int16_t tm; // <1,10>
6  int8_t derivCoeffs[5]; int16_t x[40];
7  int16_t y[40]; int16_t z[40];
8     [...]
9  while(i<NumTuples) {
10  x2 = (int64_t) (x[i] * x[i]);//<-6, 20>
11  x2 = x2 >> 10;//<-6,10>
12  y2 = (int64_t) (y[i] * y[i]);//<-2,22>
13  y2 = y2 >> 10;//<-2,11>
14  y2 = y2 << 1;//<-1,11>
15  z2 = (int64_t) (z[i] * z[i]);//<-6,20>
16  z2 = z2 >> 10;//<-6,10>
17  x2 = x2 << 5;//<-1,10>
18  y2 = y2 >> 0;//<-1,10>
19  tm = x2 + y2 +z2;  // <-1,10>
20  z2 = z2 << 5; // <-6,10>
21  magSqrt[i]=sqrt_fix(tm,-1,9,8);//<-1,10>
22  i = i + 1;}
23 while(n<NumTuples ) {
24  while(i<LpfFiltLen) {
25    if(n - i >= 0) {
26      aux0=(int64_t)(lpfCoeffs[i]
27      *magSqrt[n-i]);//<1,20>
28      aux0=aux0 >> 9; // <2,10>
29      res = res >>0; // <8,10>
30      aux0 = aux0 << 6; // <8,10>
31      res = res + aux0;} //<8,10>
32  i = i + 1;} n = n + 1;} [...]
```

Fig. 1. Left: tuned program generated with a pair |ufp, nsb| for each variable as highlighted in blue. Right: Program C generated with fixed point formats.

mental results are presented in Sect. 4. We discuss related work in Sect. 5, and conclude in Sect. 6.

2 Overview

Before we dive into the technical details of our tool, we present in this section an overview of our method using the example of a FIR low-pass filter code given in Fig. 1. The starting point of our analysis is to assume that all the variables are in a given IEEE754 precision (here we use single precision) and that a range is given for the inputs of the program. In addition, the statement `require_nsb(res,8)` is a postcondition added by the user to specify that `res` must have 8 significant bits at the end of the execution. POPiX first performs a range determination by dynamic analysis for all the program variables at each control point. Based on semantic equations, POPiX generates an ILP problem from the program source code annotated with the results of the range analysis and the accuracy requirement. This yields a system of constraints:

$$C = \left\{ \begin{array}{l} \mathsf{nsb}(+)^{\ell_{548}} >= \mathsf{nsb}(\mathtt{res})^{\ell_{549}}, \mathsf{nsb}(\mathtt{res})^{\ell_{549}} >= \mathsf{nsb}(\mathtt{if})^{\ell_{551}}, \\ \mathsf{nsb}(\mathtt{res})^{\ell_{511}} >= \mathsf{nsb}(\mathtt{res})^{\ell_{545}}, \mathsf{nsb}(\mathtt{res})^{\ell_{545}} >= \mathsf{nsb}(+)^{\ell_{548}} + 8 + carry() - 8 \\ \mathsf{nsb}(\mathtt{res})^{\ell_{545}} >= \mathsf{nsb}(+)^{\ell_{548}} + 8 + carry() - 8, \mathsf{nsb}(\mathtt{res})^{\ell_{548}} >= \mathsf{nsb}(\mathtt{res})^{\ell_{549}}, \\ \mathsf{nsb}(\mathtt{aux0})^{\ell_{547}} >= \mathsf{nsb}(+)^{\ell_{548}} + 6 + carry() - 8, \mathsf{nsb}(\mathtt{aux0})^{\ell_{543}} >= \mathsf{nsb}(\mathtt{aux0})^{\ell_{547}} \end{array} \right\} \quad (1)$$

For instance, the system C of Eq. (1) below describes the constraints generated for the addition and assignement statements for Line 23 of Fig. 1 (left hand side). Some notations can be highlighted for the system of Eq. (1). First, each variable of our program is assigned to a unique control point $\ell \in Lab$ in order to determine easily their number of significant bits. Second, the function $carry()$ is used to compute if a carry bit can occur through the operation (returns 0 or 1). Concerning scalability, we generate a linear number of constraints and variables in the size of the analyzed program (≈500 for the FIR low-pass filter code). The solution to our system of constraints gives the minimal number of bits needed with an accuracy guarantee on the results (highlighted in blue in the left hand side of Fig. 1). If we take back Line 23 under discussion, the pair $|8, 10|$ denotes that the unit in the first place of variable `res` is 8 whereas it has 10 significant bits. More details about the nature of constraints that we generate for the language of our input programs was detailed in [1].

Based on the tuning results, POPiX synthesizes the C code given in the right hand side of Fig. 1. First, it selects the best format (int16_t, int32_t, etc.) for each variable (this is called mixed-precision). For example, at lines 6 and 7, vectors `x`, `y` and `z` are defined as int16_t variables while the vector `derivCoeffs` is defined as int8_t. The data type selected by POPiX for each variable is the minimal one enabling us to encode the fixed-point value following the formats coming from the ILP solution. For example, the variable `res` has 10 significant bits (lines 29 and 31 of Fig. 1) and can consequently be encoded into int16_t. POPiX determines the initial formats $\langle M, L\rangle$ of the variables occurring in the code and synthesizes the alignments needed to change the formats, before performing some operation. For example, the shifts performed at lines 29 and 30 are done in order to align the operand of the addition of Line 31. Similarly, the shift of Line 28 is done to obtain the right format for the result of the multiplication of Line 27. Currently, the fixed-point operation are generated sequentially and some additional optimizations could be done, for example by using only one shift for lines 27 and 29.

3 Floating to Fixed-Point Programs Synthesis

POPiX workflow is based on two frameworks as depicted in Fig. 3: a developed fixed-point library (Sect. 3.1) and a precision tuning framework (Sect. 3.2). In the rest of this section, we explain how the combination of these features are achieved and which benefits they provide.

3.1 Fixed-Point Arithmetic

Since fixed-point operations rely on integer operations, computing with fixed-point numbers is highly efficient for embedded systems with small memories and simpler CPUs. However, this arithmetic is more difficult to handle for the developer. There exists some fixed-point libraries such as Libfixmath[1], Fixmath[2]

[1] https://code.google.com/archive/p/libfixmath/.
[2] http://savannah.nongnu.org/projects/fixmath/.

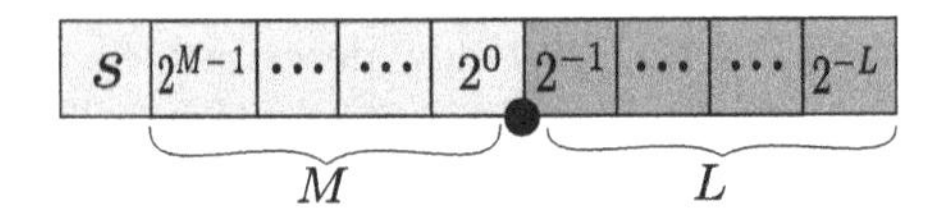

Fig. 2. Fixed-point representation of a in a format $\langle M, L \rangle$.

and FPM[3], but we have developed our own library in order to dispose of all the features we need such as mixed-precision, elementary functions, etc.

A fixed-point number is represented by a sign $s \in \{0, 1\}$, an integer value $V \in \mathbb{N}$ in base 2 and a format $\langle M, L \rangle$ as shown in Fig. 2. The number of bits before, respectively after, the radix point is $M \in \mathbb{Z}$ (respectively $L \in \mathbb{N}$). The value of a fixed-point number is obtained by multiplying the integer value V by the sign s and a scaling factor 2^{-L} as follows:

$$a = (-1)^s \times V \times 2^{-L} \ . \tag{2}$$

Example 1. *The fixed-point number* $a = 3_{<2,1>}$ *corresponds to the value 1.5. Using Eq.* (2), *we obtain* $a = (-1)^0 \times 3 \times 2^{-1} = 1.5$.

Let us note that the number of bits M of the integer part already presented in Fig. 2 is computed through the unit in the first place (ufp) defined by

$$\forall x \in \mathbb{F}, \quad \mathsf{ufp}(x) = \begin{cases} \min\{i \in \mathbb{Z} : 2^{i+1} > |x|\} = \lfloor \log_2(|x|) \rfloor & \text{if } x \neq 0, \\ 0 & \text{if } x = 0 \ . \end{cases} \tag{3}$$

Hence, the number of bits M before the radix point is given by

$$M = \mathsf{ufp}(|a|) + 1 \ . \tag{4}$$

Let W be the number of bits used to encode a. The number of bits L of the fractional part of a is

$$L = W - M - 1 \ . \tag{5}$$

The difficulty of the fixed-point representation is to manage the format $\langle M, L \rangle$ manually against the floating-point representation which manages it automatically, thanks to the exponent. Let $*$ be a fixed-point elementary operation with $* \in \{\oplus, \ominus, \otimes, \oslash\}$ and let us consider the fixed-point numbers a, b and c such that $c = a * b$. For the addition and subtraction, the resulting format of c is given by

$$\langle M^c, L^c \rangle \ = \ \langle \max(M^a, M^b) + 1, W^c - \max(M^a, M^b) - 1 \rangle \ . \tag{6}$$

For the multiplication and division, the resulting formats of c are

$$\langle M^c, L^c \rangle \ = \ \langle M^a + M^b, W^c - M^a - M^b > \ . \tag{7}$$

[3] https://github.com/MikeLankamp/fpm.

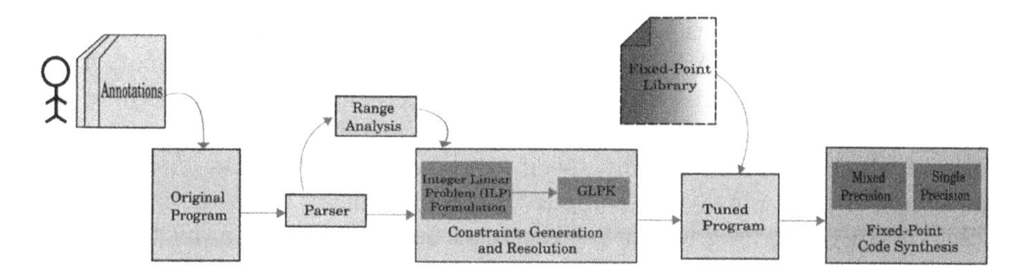

Fig. 3. POPiX workflow.

and

$$\langle M^c, L^c \rangle = \langle M^a + L^b, W^c - M^a - L^b \rangle \ . \tag{8}$$

For a fixed-point number c, the formats of $c \ll k$ and $c \gg k$ are

$$\langle M^c, L^c \rangle = \langle M^a - k, L^c + k \rangle \text{ and } \langle M^c, L^c \rangle = \langle M^a + k, L^c - k \rangle \ . \tag{9}$$

The algorithms of these elementary operations are detailed in [11,12]. Let us consider a fixed-point elementary function $f \in \{\text{abs, sqrt, sin, cos, arctan}\}$ and the fixed-point number $c = f(a)$. For the square root function, the value of the result can be approximated by the digit recurrence iteration algorithm defined in [12]. For $f \in \{\text{sin, cos, arctan}\}$, Taylor's formula is used to approximate the result. For example, the corresponding formula of sine is $\sin(a) \approx a \oplus (((a \otimes a) \otimes a) \oplus 6)$. The order of the Taylor series development depends on the number of significant bits needed for the result. In the following, we describe how we compute the fixed-point numbers occuring in our programs.

3.2 Constraint Generation by Static Analysis

POPiX presents a novel static technique based on a semantic modelling of the propagation of the numerical errors throughout the code. In practical terms, our approach depends on two integer quantities: *i*) The ufp of the values (see Eq. (3)); *ii*) A user requirement denoting the final accuracy wanted for the outputs. Hereby, the term accuracy refers to the number of significant bits required by the user on a variable of the program, denoted by nsb. Formally, let $\hat{x}$ be the approximation of x in finite precision and let $\varepsilon(x) = |x - \hat{x}|$ be the absolute error. So, if $\mathsf{nsb}(x) = k$, for $x \neq 0$, then we have

$$\varepsilon(x) \leqslant 2^{\mathsf{ufp}(x)-k+1} \ . \tag{10}$$

An ILP problem can be generated from the program source code which can be optimally solved by a LP solver (we use GLPK[4], in practice). Concerning our resulting data types, the key feature of our method consists in finding directly the minimal number of bits needed at each control point of the original program. Next, these precisions can be approximated to the upper number of bits

[4] https://www.gnu.org/software/glpk/.

corresponding to an existing format int16_t, int32_t, etc. By way of illustration, if a variable x has $\mathsf{nsb}(x) = 18$ bits, then x is tuned to the int32_t format. After solving the ILP problem, POPiX collects a new key information concerning the optimized precisions (along with the ufp integer quantity already computed by range analysis) in order to fully specify the fixed-point formats $\langle M, L\rangle$ introduced in Sect. 3.1. Finally, we call our fixed-point library to synthetize a fixed-point version of the program with only integer numbers. Through this technique, it is possible to achieve memory savings up to more than double with a precision cost that depends on the original program being optimized and energy savings up to 3.5× (see Sect. 4).

Implementation Details. POPiX has been developed in JAVA and C++ and uses the ANTLR tool v4.7.1[5] to parse the input programs. To be able to guarantee that no overflows will occur in computations, we perform a range analysis by launching the execution of the program a certain number of times in order to determine dynamically an under-approximation of the range of variables by using the ufp of the values. In the future, we plan to use a static analyzer. Nevertheless, POPiX uses the simple imperative language below.

$$x \in Id \quad \ell \in Lab \quad \odot \in \{+, -, \times, \div\} \quad math \in \{\sin, \cos, \tan, \arcsin, \log, \ldots\}$$

$$\mathbf{Expr} \ni \mathtt{e} : \mathtt{e} ::= \mathtt{c\#}p \mid x \mid e_1^{\ell_1} \odot e_2^{\ell_2} \mid math(e^{\ell_1}) \mid sqrt(e^{\ell_1})$$

$$\mathbf{Cmd} \ni \mathbf{c} : c ::= c_1^{\ell_1}; c_2^{\ell_2} \mid x = e^{\ell_1} \mid \mathtt{while}\ b^{\ell_0}\ \mathtt{do}\ c_1^{\ell_1} \mid \mathtt{if}\ b^{\ell_0}\ \mathtt{then}\ c_1^{\ell_1}\ else\ c^{\ell_2} \mid \mathtt{create\_vector}(v, s) \mid \mathtt{create\_matrix}(m, r, c) \mid \mathtt{require\_nsb}(x, n)$$

We denote by Id the set of identifiers and by Lab the set of control points of the program used to assign to each element $e \in$ **Expr** and $c \in$ **Cmd** a unique control point $\ell \in Lab$. Fortunately, POPiX is able to handle loops, conditionals and arrays. The declaration of vectors is expressed by the statement `create_vector`(v,s), while s denotes the size of the vector v. The declaration of a matrix m is expressed by the statement `create_matrix` (m, r, c), while r and c denote respectively the number of rows and columns of the matrix. The statement `require_nsb`(x,n) indicates the minimal nsb n that a variable x must have at a control point. The rest of the grammar is standard. Note that the usual mathematical elementary functions are supported.

Cost Functions. Cost functions are given as optimization objective to the linear solver. Depending on which cost function is used, different criteria may be considered for the tuning phase. POPiX currently handles the following cost functions: **CF1** Optimizes the sum of the number of significant bits of all the variables at each control point of the program. **CF2** Optimizes the sum of accuracies of only the variables assigned in the program. Compared to **CF1**, this cost function minimizes the size of the variables and the number of bits needed for the operators to store the intermediary results. Let us note that **CF2** is used in the experiments of Sect. 4. **CF3** Minimizes the maximal accuracy needed in the program, i.e. the worst accuracy at some control point of the tuned program. This function is usefull to make a program fit in a certain format (for example all variables in 16 or 32 bits.) **CF4** Optimizes only the sum of the accuracies of the

[5] https://www.antlr.org/.

arithmetic operators and elementary functions. This function is relevant from an hardware point of view, for example to limit the size of the operators in FPGAs [7]. **CF5** Minimizes the number of type conversions. Indeed, type conversions introduced by the mixed-precision tuning may slow down the execution of the programs and one may prefer a compromise between memory savings and execution time. This function addresses this problem. Note that these cost functions are modified when dealing with arrays: the tool multiplies the precision by the number of elements and this process is done only once for each array instead of several times for each use of arrays.

4 Experimental Evaluation

In this section, we conduct some experiments to show the effectiveness of our code synthesis method presented in Sect. 3.

Experimental Metrics. In our experiments, our goal is to evaluate the benefits of POPiX in terms of mixed-precision, memory savings and energy consumption which are important metrics to validate our synthesis method. We also measure the time of analysis spent by POPiX and the execution time of the fixed-point progam with respect to the floating-point program in which we assume that all variables are in single precision (32 bits) before the analysis. For our benchmarks, we use applications from FPBench[6], a synthetic benchmark for floating-point performance. We run each program with three accuracy requirements arbitrarily chosen by the user: 4, 8 and 16 bits which bound the relative error of the result. All the results we report in this section where gathered on two machines: Ubuntu 20.04 LTS, with an 2.7 GHz i7 core and 16 GB of RAM and Ubuntu 20.04 LTS, with a CPU AMD Ryzen 5 3500 u and 5.7 GB of RAM. Let us state that the reason we use the Intel machine is to exploit the Jouleit[7] tool in order to estimate the power consumption of the CPU, RAM and integrated GPU.

Results Analysis. Table 1 shows the mixed-precision configurations obtained after analysis in terms of number of variables or operations that we may tune into int8_t, int16_t and int32_t and consequently the memory savings in terms of number of bits. The second left-most column of Table 1 headed "call" refers to the number of elementary functions in the code and the next column headed "op" denotes the number of elementary operations. In this experiment, we assume that 100% is the percentage of all variables initially in single precision. Clearly, the memory savings compared to the initial number of bits for the majority of the original programs is considerable reaching 75% for "CRadius" program for a requirement of 4 bits. For instance, the "carbonGas" program has a total of 13 variables all in single precision before analysis. In the synthesized code we obtain 7 variables tuned into int8_t and 6 variables in int16_t achieving a gain in number of bits of 63,5%. Concerning "azimuth" program, our analysis failed

[6] https://fpbench.org/.
[7] https://github.com/powerapi-ng/jouleit.

Table 1. Mixed fixed-point formats in 4, 8 and 16 bits for the synthesized program with the precentages of the number of bits saved.

			4 bits				8 bits				16 bits			
Program	call	op	8	16	32	%	8	16	32	%	8	16	32	%
azimuth	7	7	1	0	17	**4.2**	0	1	17	**2.8**	-	-	-	-
carbonGas	0	7	7	6	0	**63.5**	2	11	0	**53.8**	1	1	11	**9.6**
CRadius	1	3	6	0	0	**75.0**	0	6	0	**50.0**	0	0	6	**0.0**
CTheta	1	3	4	0	3	**42.9**	0	4	3	**28.6**	0	0	7	**0.0**
doppler1	0	7	9	1	0	**72.5**	1	9	0	**52.5**	0	1	9	**5.0**
doppler2	0	7	9	1	0	**72.5**	3	7	0	**57.5**	0	3	7	**15.0**
doppler3	0	7	9	1	0	**72.5**	2	8	0	**55.0**	0	2	8	**10.0**
instantCurrent	3	18	7	14	7	**43.8**	3	18	7	**40.2**	0	3	25	**5.4**
jetEngine	0	29	7	18	5	**47.5**	3	15	12	**32.5**	0	3	27	**5.0**
LeadLagSystem	1	17	2	29	2	**48.5**	0	31	2	**47.0**	0	0	33	**0.0**
LowPassFilter	0	0	0	330	0	**50.0**	0	330	0	**50.0**	0	4	326	**0.6**
CX	1	3	2	0	4	**25.0**	0	2	4	**16.7**	0	0	6	**0.0**
CY	1	3	2	0	4	**25.0**	0	2	4	**16.7**	0	0	6	**0.0**
triangle12	1	9	7	6	0	**63.5**	0	12	1	**46.2**	0	0	13	**0.0**
turbine1	0	14	4	13	0	**55.9**	0	17	0	**50.0**	0	0	17	**0.0**
turbine2	0	10	10	3	0	**69.2**	0	13	0	**50.0**	0	0	13	**0.0**
turbine3	0	14	3	14	0	**54.4**	0	17	0	**50.0**	0	0	17	**0.0**

to infer mixed precision for a user accuracy requirement of 16 bits whereas it reaches 4.2% and 2.8% respectively for requirements of 4 bits and 8 bits. Another observation is that for a requirement of only 4 bits, 17 variables are tuned into int32_t. A possible explanation of this result is the call to the elementary functions in the code (call = 7) which can use intermediate variables that request greater precision than the user accuracy requirement.

Figure 4 depicts the energy consumed by the execution of the benchmarks for a requirement of 4 bits. We observe that the fixed-point version codes require significantly less energy than the floating-point codes. For instance, the energy saved on CPU and DRAM reaches ≈ 43% for "carbonGas" program and more than 75% for "jetEngine" program. Let us note that this observation is also valid for the remaining user requirements with slight variations of savings. Finally, we present the results in terms of speed for each of our benchmarks in Table 2. We denote respectively by "t_{float}", "$t_{synthesis}$" and "t_{fix}" the time of execution of floating-point programs, the total synthesis time of POPiX and the execution time of fixed-point programs all given in milliseconds. We visualize that the time spent by POPiX for the majority of benchmarks is negligible not exceeding 342 ms for the "carbonGas" program (≈ 30 LOCs). Although our synthesis method

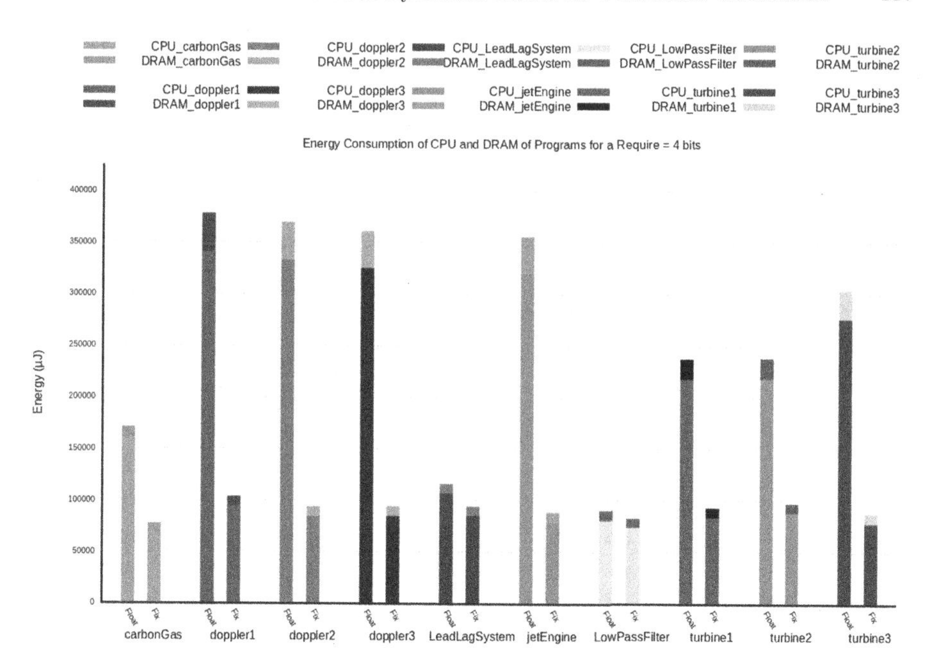

Fig. 4. Measurement of the energy consumption (CPU and DRAM) of the floating-point and fixed-point version of our benchmarks.

is fast (few seconds), we observe that the execution time remains the same for the floating and fixed-point codes with negilible slow-down for some benchmarks.

5 Related Work

In recent years, many authors have investigated the possibility of automating fixed-point code synthesis. A similar approach to our work is the TAFFO tool proposed by Cherubin et al. [4]. TAFFO is a static precision tuning tool that converts floating-point computations into a fixed-point version with comparable semantics. The common point between POPiX and TAFFO is that the estimation of the errors is generated by the precision tuning process. Meanwhile, POPiX is much faster and takes only few seconds to synthesize the new fixed-point formats of the program.

Another solution for code conversion was introduced by Cattaneo et al. [3]. Their method relies on a self-contained compiler transformation pass implemented within LLVM to perform the conversion. Their tool was especially dedicated to MIOSIX, a real time operating system targeting embedded system.

The goal of the dissertation of Jha [9] is to give an algorithm for optimal fixed-point expressions synthesis based on inductive synthesis. Two years later, Darulova et al. [5] proposed a fixed-point program synthesis methodology based on expression rewriting and genetic programming. Their algorithm uses

Table 2. Execution time measurements obtained during the experiments.

		4 bits		8 bits		16 bits	
Program	t_{float}	$t_{synthesis}$	t_{fix}	$t_{synthesis}$	t_{fix}	$t_{synthesis}$	t_{fix}
azimuth	1.91	340	1.8	301	2.6	233	3.2
carbonGas	0.17	342	0.29	233	0.31	201	0.46
CRadius	0.10	240	1.31	192	1.33	186	1.36
CTheta	0.38	203	0.56	223	0.57	273	0.57
doppler1	0.16	205	0.53	249	0.54	225	0.57
doppler2	0.24	188	0.28	225	0.29	207	0.30
doppler3	0.17	243	0.30	195	0.32	207	0.34
instantCurrent	1.01	264	1.99	265	2.36	273	2.73
jetEngine	0.34	294	1.18	264	1.24	276	1.82
LeadLagSystem	0.53	352	0.71	394	0.86	346	1.09
CX	0.43	178	0.29	208	0.37	175	0.52
CY	0.39	203	0.41	197	0.55	197	0.68
triangle12	0.17	199	1.57	208	1.63	203	2.53
turbine1	0.24	220	0.31	228	0.35	269	0.49
turbine2	0.28	246	0.49	222	0.61	212	1.01
turbine3	0.48	247	0.47	253	0.49	217	0.83

abstract interpretation to estimate the error bound of a fixed-point implementation. However, the latter two techniques provide pessimistic bounds for non-linear expressions and are limited to straight-line programs. Coversely, Aslan et al. [2] developed a tool that takes an n-bit fixed-point input and creates an m-bit floating-point output with IEEE754 and custom formats.

In the context of polynomials, linear filters and signal processing algorithms, the members of the DEFIS project [13] presented many approaches for fixed-point code synthesis. To mention a few, the idea described in [6] works by inferring high-level convolution operations from the original source code, and modeling them as part of the program representation. In addition, Najahi et al. [12] presented an automated approach to synthesize codes in fixed-point arithmetic for some linear algebra basic blocks. They take a mathematical description of the problem as well as the range of the input variables and generate fixed-point code. Lopez [11] addresses the transformation of linear filters and controllers into hardware operators using fixed-point arithmetic. His main contribution is a complete error analysis, with respect to the internal word-lengths and the formulation of the word-length optimization as a convex non-linear integer optimization problem solved using appropriate heuristics. An extension of this work to the full class of linear time invariant algorithms has been proposed in [14].

6 Conclusion

In this article, we have presented a new method for fixed-point code synthesis respecting an accuracy requirement imposed by the user. Our method is based on a static analysis of the code implemented by means of system of constraints which gives the minimal format needed to encode each value. Experimental results show the performance of the codes synthesized in terms of execution time, memory and energy savings on a set of benchmark related to embedded systems.

In the future we would like to validate our method by considering architectures more commonly used in embedded systems. Also, we aim at generating hardware instead of software fixed-point implementations, using FPGAs. Targeting FPGAs has two justifications: this type of hardware is becoming more and more popular today and it presents the advantage of allowing fully custom designs. Finally, for adoption reasons in real-world applications, we aim at extending POPiX in order to handle full C programs via an integration to LLVM.

References

1. Adjé, A., Ben Khalifa, D., Martel, M.: Fast and efficient bit-level precision tuning. In: Drăgoi, C., Mukherjee, S., Namjoshi, K. (eds.) SAS 2021. LNCS, vol. 12913, pp. 1–24. Springer, Cham (2021). https://doi.org/10.1007/978-3-030-88806-0_1
2. Aslan, S., Oruklu, E., Saniie, J.: A high-level synthesis and verification tool for fixed to floating point conversion. In: 55th IEEE International Midwest Symposium on Circuits and Systems, MWSCAS, pp. 908–911. IEEE (2012)
3. Cattaneo, D., Di Bello, A., Cherubin, S., Terraneo, F., Agosta, G.: Embedded operating system optimization through floating to fixed point compiler transformation. In: 21st Euromicro Conference on Digital System Design, DSD, pp. 172–176. IEEE Computer Society (2018)
4. Cherubin, S., Cattaneo, D., Chiari, M., Agosta, G.: Dynamic precision autotuning with TAFFO. ACM Trans. Archit. Code Optim. **17**(2), 10:1–10:26 (2020)
5. Darulova, E., Kuncak, V., Majumdar, R., Saha, I.: Synthesis of fixed-point programs. In: Proceedings of the International Conference on Embedded Software, EMSOFT, pp. 22:1–22:10. IEEE (2013)
6. Deest, G., Yuki, T., Sentieys, O., Derrien, S.: Toward scalable source level accuracy analysis for floating-point to fixed-point conversion. In: The IEEE/ACM International Conference on Computer-Aided Design, ICCAD, pp. 726–733. IEEE (2014)
7. Gao, X., Constantinides, G.A.: Numerical program optimization for high-level synthesis. In: Constantinides, G.A., Chen, D. (eds.), Proceedings of the 2015 ACM/SIGDA International Symposium on Field-Programmable Gate Arrays, pp. 210–213. ACM (2015)
8. Gupta, S., Agrawal, A., Gopalakrishnan, K., Narayanan, P.: Deep learning with limited numerical precision. In: Proceedings of the 32nd International Conference on International Conference on Machine Learning, vol. 37, ICML 2015, pp. 1737–1746. JMLR.org (2015)
9. Jha, S.: Towards Automated System Synthesis Using SCIDUCTION. PhD thesis, University of California, Berkeley, USA (2011)

10. Lin, D., Talathi, S. and Annapureddy, S.: Fixed point quantization of deep convolutional networks. In: Proceedings of the 33rd International Conference on International Conference on Machine Learning - Volume 48, ICML 2016, pp. 2849–2858. JMLR.org (2016)
11. Lopez, B.: Implémentation optimale de filtres linéaires en arithmétique virgule fixe. (Optimal implementation of linear filters in fixed-point arithmetic). PhD thesis, Pierre and Marie Curie University, Paris, France (2014)
12. Martel, M., Najahi, A., Revy, G.: Trade-offs of certified fixed-point code synthesis for linear algebra basic blocks. J. Syst. Archit. **76**, 133–148 (2017)
13. Ménard, D.: Design of fixed-point embedded systems (DEFIS) French ANR project. In: Design and Architectures for Signal and Image Processing, DASIP, pp. 1–2. IEEE (2012)
14. Volkova, A.: Towards reliable implementation of digital filters. (Vers une implémentation fiable des filtres numériques). PhD thesis, Pierre and Marie Curie University, Paris, France (2017)

Optimized Hardware and Software Implementations for Image Processing and Health Applications

Data-Type Assessment for Real-Time Hyperspectral Classification in Medical Imaging

Manuel Villa[1(✉)], Jaime Sancho[1], Guillermo Vazquez[1], Gonzalo Rosa[1], Gemma Urbanos[1], Alberto Martin-Perez[1], Pallab Sutradhar[1], Rubén Salvador[2], Miguel Chavarrías[1], Alfonso Lagares[3], Eduardo Juarez[1], and César Sanz[1]

[1] Universidad Politécnica de Madrid (UPM), 28040 Madrid, Spain
{manuel.villa.romero,jaime.sancho,guillermo.vazquez.valle, guillermo.vazquez.valle,gonzalo.rosa.olmeda,gemma.urbanos, a.martinp,pallab.sutradhar,miguel.chavarrias, eduardo.juarez,cesar.sanz}@upm.es

[2] CentraleSupélec, CNRS, IETR UMR 6164, 35576 Rennes, France
ruben.salvador@centralesupelec.fr

[3] Instituto de Investigación Sanitaria Hospital 12 de Octubre (imas12), 28041 Madrid, Spain
alfonso.lagares@salud.madrid.org

Abstract. Real-time constraints in image processing applications often force their optimization using hardware accelerators. This is the case for intraoperative medical images used during surgical procedures. In this context, the challenge consists in processing large volumes of data while employing high complexity algorithms in a limited period of time. Newly developed algorithms must meet both quality-accurate and hardware-efficient characteristics. In this work, we have evaluated the impact of using different data types in a processing chain to classify tissues from hyperspectral video in surgical environments. The software was run on two different embedded CPU+GPU platforms. The results show an improvement in performance by up to 9 times without increasing power consumption by reducing the bit depth from 64 to 16. The impact these reduction have on quality has been measured analytically, by calculating the RMSE, and subjectively, by surveying neurosurgeons. In both cases the results show a minimal impact on the overall quality.

Keywords: HSI · ML · tumor · video · embedded · GPU · real-time

1 Introduction

The use of different types of imaging techniques has proven its usefulness in medicine since the first use of X-rays. Today, both image sources and image

This work was supported by the Regional Government of Madrid (Spain) through NEMESIS-3D-CM project (Y2018/BIO-4826).

K. Desnos and S. Pertuz (Eds.): DASIP 2022, LNCS 13425, pp. 123–135, 2022.
https://doi.org/10.1007/978-3-031-12748-9_10

processing techniques continue evolving. These technologies are now combined with machine learning (ML) techniques and immersive reality offering clinicians useful tools for both diagnosis and treatment of a wide range of pathologies. However, the inherent complexity of increasingly refined processing algorithms makes real-time computing a challenge in this field of application. Moreover, space constraints, power consumption and hardware performance must be taken into account in controlled healthcare environments.

With regard to the use of diagnosis imaging, hyperspectral (HS) imaging (HSI) has proven to be useful in assisting in the differentiation between different types of tissues [1]. This is possible since HSI allows to establish correlations between the spectral features and the materials captured in the scene. An extensive review about HSI usage in medicine was done by G. Lu and B. Fei in [2]. In the field of neurosurgery, predictive models based on ML techniques can be generated from the information provided by HS cameras [3]. Recently, new faster snapshot HS sensors enable capturing HS video (HSV), opening the path to HSV-classification systems [4]. For this reason, the design of the processing pipelines must be highly efficient, making the best use of the available hardware resources while not degrading, in an appreciable way, the quality of the images presented to the healthcare professionals. Existing works in the scientific literature often offer performances far from real-time [5].

The other battlefront therefore lies in the use that complex processing algorithms require of the available hardware. In terms of both the implementation of ML algorithms and image processing, graphics processing units (GPU) are showing an almost unassailable market leadership. In [6], J. Sancho et al. presented a GPU-accelerated multiview hyperspectral depth estimation tool for medical imaging. In [7], a survey highlighting the impact of the GPUs in the magnetic resonance images processing techniques and reconstruction is outlined.

In this work, we propose to make an efficient use of the hardware resources, using heterogeneous CPU+GPU based embedded systems as target platforms, by means of an effective software implementation. Specifically, an implementation of a complete HSV processing and classification suite based on the use of half, float and double types has been made with a Support Vector Machine (SVM) classifier. A specific data type is employed for the whole algorithm chain, as combinations of different data types are not expected to improve the quality or processing time. The test bench is composed of a set of HS videos captured in real environments, i.e. a neurosurgery operating theatre. The results are grouped to show the differences obtained in terms of accuracy, energy consumption, processing time and classification differences. Additionally, a survey on the subjected quality of each solution has been conducted.

2 Background

In Sect. 3, the proposed HSV processing chain is depicted. It has been improved from the one presented in [4]. The background on which some of the blocks that originally make up this chain are based is detailed below:

1. Pre-processing chain: this stage is intended for generating a HS cube from the raw snapshot capture. It begins with the cube conformation, a new arrangement of the pixels depending on the filter pattern of the snapshot camera. Then, it is calibrated using a white and black reference; in this way, the pixel values are transformed to a percentage of reflectance. This calibrated cube also needs a spectral correction that correct the artifacts introduced by the spectral filter. Finally, this cube is normalized in order to adjust the light energy received by the sensor.
2. Classification chain: although the purpose of this work is not the implementation of a classification model, this represents a fundamental part of the proposed processing suite. To this end, the classification model and training process used is based on G. Urbanos et al. work [3]. Where three ML-based algorithms, Random Forest, SVM and Convolutional Neural Networks, are compared regarding the accuracy when classifying brain tissues. It is worth mentioning that while the classification results are promising, in some cases the accuracy obtained slightly exceeds 60% for some tissues. This limited result logically carries over to this solution. SVM classifier is essentially integrated by several matrix-vector multiplications, which can be easily GPU-accelerated.

3 Algorithm and Acceleration

The HSV-classification algorithm chain employed in this work is presented in Fig. 1. As depicted, this is divided into two main stages, (i) a pre-processing chain intended to generate a proper HS cube from the HS raw images and (ii) a classification algorithm along with some additional processes.

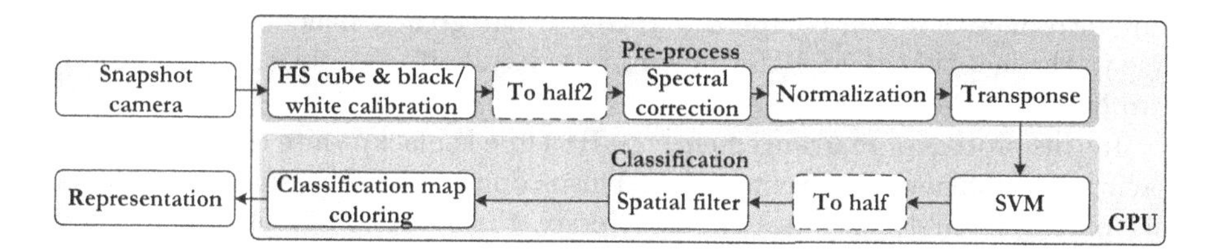

Fig. 1. Algorithm chain. Boxes with dashed-lines represent stages only performed in the half implementation of the algorithm chain.

This algorithm chain has been implemented and accelerated in two CPU+GPU embedded platforms (see Sect. 4.1) and using three different data types: double (64-bit floating point), float (32-bit floating point) and half (16-bit floating point). This allows the analysis of three different scenarios where the numerical accuracy is gradually lowered while improving the time-energy performance. Due to the specific architecture of the GPUs employed, the acceleration of float and double is similar, whilst the acceleration using half is significantly different. All the stages are further explained in the following subsections. As a general remark, all the computation was performed in the GPU using the corresponding data type, whilst in the CPU only the acquisition and the representation is performed, always using an unsigned 8-bit integer.

3.1 Pre-processing Chain

This chain begins with the conformation of the HS cube from a raw capture and its black/white calibration. The HS snapshot camera employed in this work produces a video streaming of 2045×1085@170 FPS raw captures. These captures present a 5×5 pattern filter replicated in the image, meaning that the 25 pixels included within these pattern filters are representing 25 different wavelengths. For this reason, the HS cubes produced by this camera are reduced to 409×217 spatially, and 25 bands spectrally. However, in order to fit the warp size of the GPU (32 threads) and optimize the processing, these bands are padded to 32, including zeros in the last 7 positions.

This rearrangement process is accelerated in the GPU in a kernel that makes use of a single thread for each pixel in the raw HS image, meaning that every thread is copying its HS raw value to a new memory position, given an index transformation. The method chosen to store the HS cube is denominated band interleaved pixel (BIP), given its benefits in the following processing stages. In this method, all the bands of a HS pixel are located contiguous, and the following HS pixel begins at its end. This is presented in Fig. 2, with differences between double/float and half. In the former case, every memory position holds a single value, and the information is stored as introduced before. In the later case, half, every memory position holds an structure of two half values, denominated half2 (accessed as .x and .y, respectively). In this way, the number of half2 is halved compared to the double/float implementation. Although this data structure introduces a higher complexity, the GPU can greatly benefit from it. This is due to the fact that the GPU is able to process a half2 as a float, i.e., to process two half with the same time-energy cost as a single float.

In this process of rearrangement, the HS cube is black/white corrected before saving it in the new memory position. This is done, as said before, with a thread per pixel that first calculates the new index, then correct the pixel and finally save it to the new position. The calibration is performed using a black calibration image and a white calibration image following this formula:

$$HS'(x, y, \lambda) = \frac{HS(x, y, \lambda) - B(x, y, \lambda)}{W(x, y, \lambda) - B(x, y, \lambda)} \tag{1}$$

The next stage is the spectral correction, intended to improve the spectral quality of the HS cube by reducing the effect of the filter side undesirable spectral lobes. To do so, all the bands for each HS pixel are multiplied by a correction matrix that produces the corrected value for every band. This is a vector by matrix multiplication formulated as:

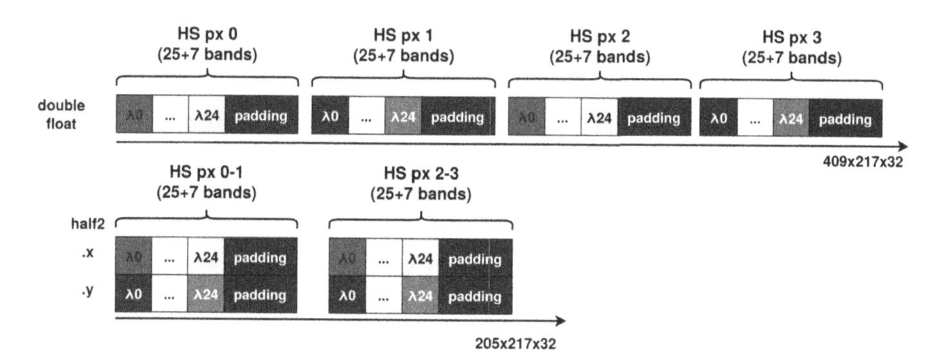

Fig. 2. HS cube in memory. Lighter boxes refer to even HS pixels, darker boxes refer to odd HS pixels.

$$HS'(x, y, \lambda) = \sum_{i=0}^{\Lambda} HS(x, y, i) * C(i, \lambda) \tag{2}$$

C refers to the correction matrix and Λ refers to the number of bands, 32.

This stage is accelerated in a kernel that performs a warp multiplication in every HS pixel. This is possible thanks to the padding to 32 bands explained before and the BIP format used. With a thread addressing every HS cube position, this allows (i) the coalesced access of memory, as 32 contiguous memory positions are always read by 32 contiguous threads and (ii) the use of shuffle warp operations, to share register data between threads in the same warp.

This process begins with a coalesced data load from every thread, which means that every thread in a warp hold in a register a band from a HS pixel. Then, the values in the 32 threads (that are shared between all of them) are multiplied by the corresponding correction matrix vector, depending on the index of the thread (that is the band number). This is repeated 32 times for every thread so that the multiplication of a band by the correct factor is accumulated in a thread register. Once this finishes, every thread in the warp has the corrected value in that register, being able to store it with a coalesced write. In addition, as the correction matrix is small enough (32×32), it is cached and quickly accessed by every thread. This process is summarized in Fig. 3 for the first pixel and warp. When using half numbers, the process is similar, with the difference that a warp is correcting two HS pixels with the same number of operations as before.

The following stage is the normalization, where all the bands of an HS pixel are normalized by the mean squared value of all its bands. This is expressed as:

$$M(x, y) = \frac{1}{\Lambda}\sqrt{\sum_{i=0}^{\Lambda} HS(x, y, i)^2} \longrightarrow HS'(x, y, \lambda) = \frac{HS(x, y, \lambda)}{M(x, y)} \tag{3}$$

This operation is performed in a kernel that makes use of the warp padding as well. In this case, the benefits are similar as in the previous stage; memory

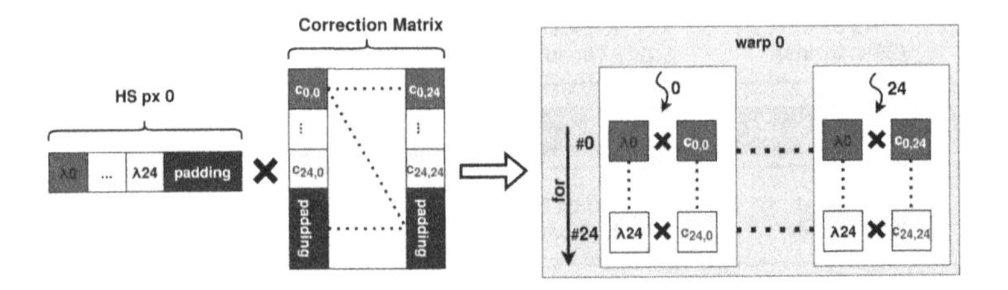

Fig. 3. Spectral correction stage.

accesses are coalesced and warp shuffle operations allow sharing information within the warp without the need of an explicit synchronization. This is employed to, using a thread per pixel, first, load an HS pixel to a warp (32 threads read 32 bands), then, perform a warp reduction [8] to obtain the mean squared value and finally, share this value to all the threads, which divide its initial pixel value to the mean.

The last stage in the pre-processing is the transpose, needed to improve the performance of the operations in the classification chain, given the low number of classifiers/classes that would underuse the warp occupancy (compared to the 25 HS bands). The operation, performed in a kernel, is a memory rearrangement between the previous format, BIP and the new format, band sequential (BSQ), where all the pixels of a band are contiguous and bands appear sequentially. As before, half values are grouped in containers of two, a half2; however, they are also transformed to BSQ format. In this case, the band padding is not longer needed but the rows need to be padded to an even number in order to maintain the half2 structure. The new memory format is represented in Fig. 4.

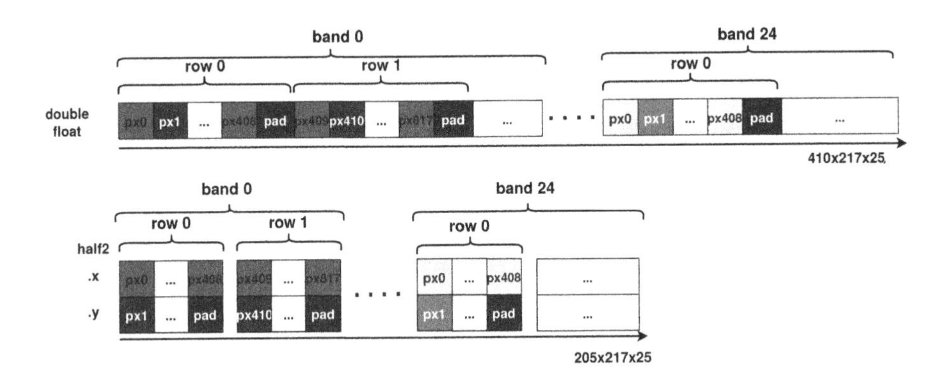

Fig. 4. Transpose stage.

3.2 Classification Chain

The first stage in the classification chain is SVM. This process is divided in two different kernels: (i) score calculation, where the distance between every HS pixel and the SVM hyperplanes from the model is calculated and (ii) probability estimation, where using the distances, the probability of belonging to a class for every HS pixel is estimated.

The score calculation kernel performs a matrix multiplication between the HS pixel (with 25 bands) and all the SVM hyperplanes from the model in order to get a distance per hyperplane. This is accelerated using a bi-dimensional grid where the x dimension represents the linear spatial dimensions of the cube (409×217) and the y dimension represents the number of classifiers in the model. In this kernel, every thread in x, y calculates a dot product between a HS pixel (x) and a classifier (y) by means of a for loop that iterates through all the bands in the HS pixel and the classifiers hyperplane. Finally, an offset is added to this value. This is similar for the half version, which process two distances per thread. The result of this process is a cube with the same spatial dimensions and a number of bands equal to the number of classifiers/hyperplanes. Figure 5 depicts this process.

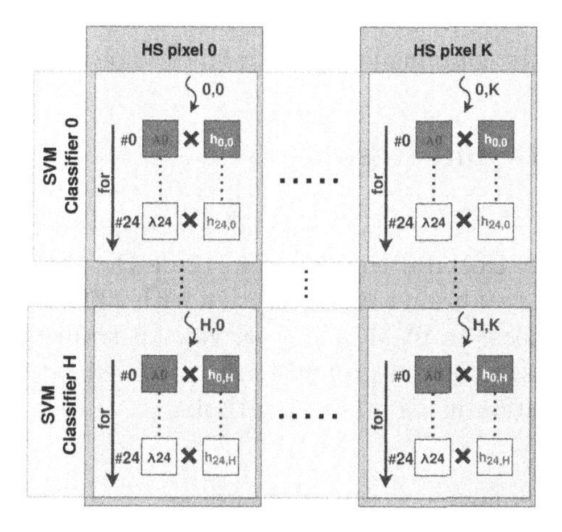

Fig. 5. SVM stage. H refers to the maximum number of classifiers and K refers to the number of spatial pixels (409×217).

The probability estimation kernel employs the sigmoid function to estimate the probability of each class (not classifier/hyperplane) using the distances to the hyperplanes. To do so, a bi-dimensional grid is employed. This addresses a thread per HS pixel (cube spatial dimensions) in the x dimension and per class in the y dimension. In this way, every thread calculates the probability of belonging to its class (y index) using all the distances previously calculated. As before, this is a per-pixel operation that is performed per two pixels in the half implementation. The result of this stage is a cube with the same spatial dimensions and a number of bands equals to the number of classes.

In the next step, only in the half implementation, the probabilities from SVM are transformed from half2 to half data containers. This means that the memory structure at this point is equal to double/float, only changing the number of bits per pixel. This is needed to perform the last part of the classification chain, the spatial filter and coloring, whose performance is increased when the data in memory has a real spatial ordering. The spatial filter performs a 3×3 window gaussian filtering in every probability class map. This is done with a kernel addressing a thread for each spatial dimensions of the cube that performs an average value in the window. This is repeated for every class within the threads and then the value is stored with the same structure.

Finally, this probabilities are converted to a color map with the last kernel. This kernel works in two different modes (i) color using probabilities or (ii) color for the max class. In the first one, a linear combination is performed using the probability per class. In this way, the color of a pixel is a mixed color depending on the probability and color label of each class. The second one only assigns the color label of class with the highest probability. The two modes are used in two scenarios: (i) subjective analysis, as mixed colours are more informative for neurosurgeons (some parts of tumor edges may have less probability to be cancer than an inner one) and (ii) objective comparison against the ground-truth, as the ground-truth only contains pixels with labels, not a mix between them.

4 Experiments and Results

In order to test the processing chain, a set of 6 videos, obtained during 4 different patient operations at Hospital Universitario 12 de Octubre (HU12O) in Madrid (Spain), capturing a real brain tumor resection with 200 frames each is employed. To classify those videos, an SVM model per video is trained using the remaining 5 videos. These videos have a ground-truth provided by the neurosurgeon in charge of that operation using a labelling tool.

4.1 Platforms Features and Experiments

Two embedded platforms integrated by heterogeneous, General Purpose Processor (GPP) and GPU, System-on-Chip (SoC) have been used to conduct the experiments. The first one is a Jetson AGX Xavier platform and integrates 4 processing clusters, each one with 2 ARM cores running a maximum clock of 2.26 GHz. Each pack of two cores share 2 MB of L2 cache memory and have access to 4 MB of L3 cache memory. This platform has 512 GPU cores with a boost frequency of 1.37 GHz and 64 Tensor cores. The second one is a Jetson Nano , and integrates 4 ARM cores with a maximum clock of 1.43 GHz alongside 128 GPU cores with a maximum operating frequency of 921 MHz and 2 MB of unified cache memory.

These platforms have different energy consumption profiles [9]. For this work, two profiles have been selected for each one; (i) the one with the highest performance and with the highest power consumption, denominated *max*, and (ii) the one with the lowest performance and lowest power consumption, denominated *min*. For Jetson Xavier, *max* and *min* are set up to a maximum power consumption of 30 W and 10 W, respectively. For Jetson Nano, *max* and *min* are set up to 10 W and 5 W, respectively. Therefore, the experiments conducted are the following:

(i) The whole processing chain has been run in the two platforms and with the two different profiles.
(ii) For each platform and profile, three implementations of the processing chain have been executed: using double, float or half data types.
(iii) The resultant video classification map is generated while the processing time and the instantaneous power consumption is measured for the twelve combinations (2 platforms, 2 profiles and 3 data types).
(iv) All the videos are compared to the ground-truth in order to get an accuracy result (focused to test numerical differences between data types). In this work only the comparison for data types is presented (there are not differences between platforms nor profiles). It is important to remark that only the pixels of ground-truth are tested, not the whole image.
(v) The videos with different data types are tested in terms of numerical precision. To do so, the classification maps are scaled by a color label depending on its SVM probability (see Sect. 3.2) and then they are compared in pairs using root mean square error (RMSE) (double-float, float-half, double-half). In this experiment the whole frame is compared.
(vi) The videos processed with different data types are presented in a survey of 10 cases to the neurosurgeon service of HU12O, to measure the subjective differences and the medical impact between them. The survey was designed following ITU-T P.800 recommendation and comparison category rating methodology. In addition, 20% were control sequences. Neurosurgeons were asked to rate different data type pairs of videos from -3 to 3.

4.2 Objective Results

The average of a 5-round execution frames per second (FPS) achieved for each platform, profile and data type are depicted in the left side of Fig. 6. This chart shows the differences in FPS between data type for every platform/profile combination. Lowest confidence levels are also provided, highest levels slightly outperform the average. As expected, the processing time for each data type is proportional to the number of bits employed. However, the proportion is not always a factor of two; in the case of both platforms and profiles (with slight differences), the float implementation performs around 5 times better than double and the half implementation performs around 1.6 times better than the float one. Comparing the double/float implementations, they use exactly the same number of threads in every kernel, but processing two times the number of bits. This entails doubling the time accessing memory and the use of fp64 units, which in this architecture is halved compared to fp32 units. In addition, in kernels where register memory plays an important role the differences are even more noticeable, resulting in an average of 5 times slower (a 64-bit register is the composition of two 32-bit registers, slowing shuffle warp operations). Comparing float/half implementations, in half, the number of threads is halved, accessing every thread two half values (16+16 bits) and using a single fp32 unit to compute them. The result is as processing the half number of threads, in terms of computation. However, this only occurs to several kernels and there is a need to convert from half to half2 and vice-versa, lowering the speed-up factor to around 1.6. The comparison between platforms and profiles shows that the difference between profiles within Jetson Nano (Nano) is slight, as only the GPU frequency is increased. However, in Jetson Xavier (XV), both GPU and memory frequency are increased, showing a high increment in the number of FPS for every data type. Finally, the differences between platforms come from four factors: (i) the memory bandwidth is doubled, (ii) the memory/GPU frequency is higher, (iii) the number of functional units is increased and (iv) the microarchitecture is different. This complex scenario makes difficult the comparison.

In the right side of Fig. 6, the average power demand in Watts per case is also shown. It can be observed significant differences, almost $\times 2$ between *max* and *min* profiles over Nano platform and $\times 3$ in the case of the Xavier. Moreover, the results show how the use of different data types may affect the power demand by extra up to 10%. Comparing platforms, it is interesting to notice that Jetson Xavier can utilize the same average power as a Jetson Nano (given the appropriate profiles) while achieving a performance gain.

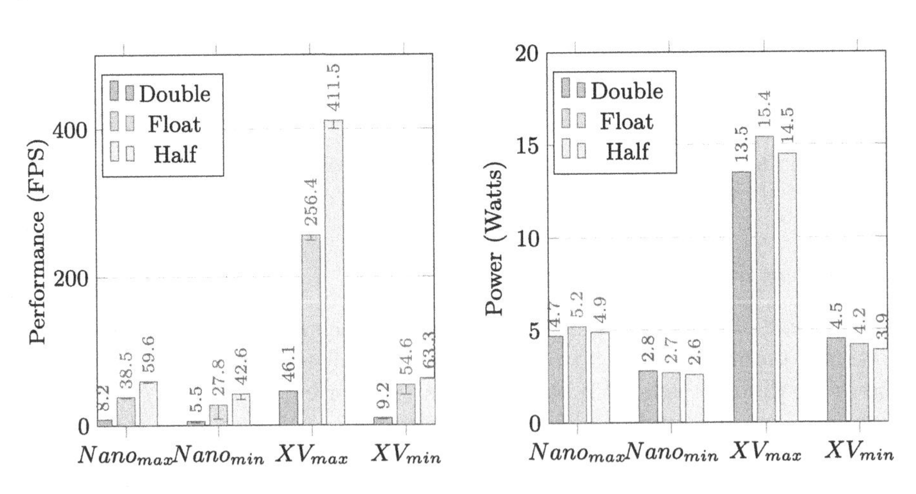

Fig. 6. Performance measured in FPS (left) and average power demand in watts (right).

In Table 1, the objective quality results are presented. This shows the differences between data types using RMSE (8-bit precision) as metric and also the accuracy (number of correct pixels), compared to the ground-truth. As can be seen, the difference between double-float is near 0 and the one between double/float-half is barely 1, exposing that the classification maps are almost equal for every data type. A similar result is extracted from the accuracy values, which are almost the same for the three cases.

Table 1. Objective quality comparison. RMSE per data-type and accuracy compared to the ground-truth obtained for each implementation.

Data type	RMSE			Accuracy
	Double	Float	Half	
Double	0	0.007	1.115	64.21%
Float	-	0	1.116	64.21%
Half	-	-	0	63.89%

4.3 Subjective Results

The result of the proposed processing chain is a classified map painted with colors the different kind of tissue captured in the scene. Figure 7 shows respective frames of the resulting classification map comparing float-half and double-half processing suits. As it can be observed, the differences are limited to a few pixels barely noticeable after enlarging the image.

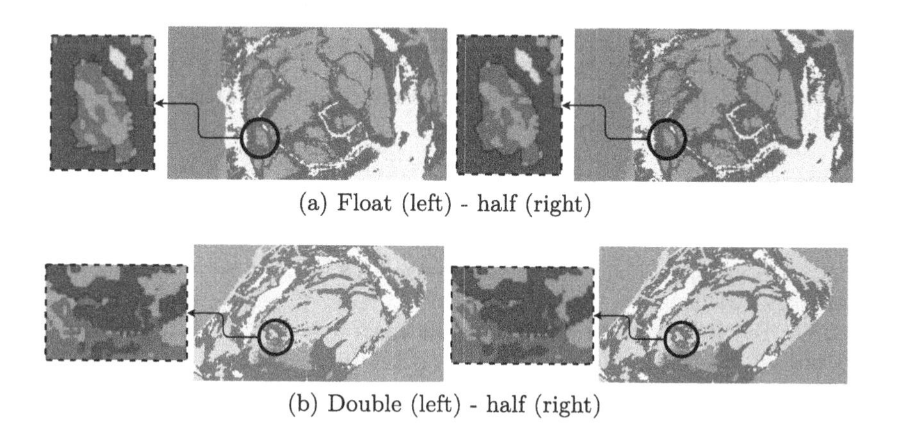

(a) Float (left) - half (right)

(b) Double (left) - half (right)

Fig. 7. Captured frame of a classification map, using different data type processing chains. An enlarged region of interest (ROI) is provided.

Finally, a total of 10 surveys, totaling 76 valid comparisons between videos were completed by neurosurgeons following the methodology described in

Sect. 4.1 *vi*. 79% of the responses indicated that there was no difference between videos processed with the half chain versus those processed with both double and float. 12% of responses indicated that videos processed with double or float were subjectively better than those processed with half.

5 Conclusions and Future Lines

In this work, a HSV-classification processing suite for use in surgical environments, optimized for embedded CPU+GPU platforms, has been presented. It has been assessed in terms of performance, measured as FPS, and power consumption, given three different data type implementations: double, float and half. Two platforms have been used for the experiments: a high-performance platform, Jetson Xavier, where it is observed that the half implementation achieved 8.9 times the performance of double type while maintaining the same power consumption. This trend has been verified on the low performance platform, Jetson Nano, where the half-based implementation offered 7.7 times higher performance than double, while reducing energy consumption a 8%. The different data type results were assessed using RMSE and the accuracy compared to a ground-truth. A maximum degradation of 1.116 units (in a scale of 8-bits precision) and less than 0.5% in accuracy was observed when using half instead of float or double data types, showing a slight difference between the classification map for the different data types. Subjectively, the differences are hardly appreciable. Additionally, a subjective survey has been conducted over a significant number of neurosurgeons, showing no subjective differences between data types.

Further work will include more powerful and accurate classification models in the processing chain. In addition, it is expected to implement and accelerate other algorithms while conserving the real-time video constraint. Moreover, a more detailed analysis of the results observed between the platform hardware features and its impact on the objective results will be conducted.

References

1. Fabelo, H., et al.: An intraoperative visualization system using hyperspectral imaging to aid in brain tumor delineation. Sensors **18**, 430 (2018). https://doi.org/10.3390/s18020430
2. Lu, G., Fei, B.: Medical hyperspectral imaging: a review. J. Biomed. Opt. **19**(1) 010901 (2014). https://doi.org/10.1117/1.JBO.19.1.010901
3. Urbanos, G., et al.: Supervised machine learning methods and hyperspectral imaging techniques jointly applied for brain cancer classification. Sensors **21**(11), 3827 (2021). https://doi.org/10.3390/s21113827
4. Sancho, J., et al.: An embedded GPU accelerated hyperspectral video classification system in real-time. In: XXXVI Conference on Design of Circuits and Integrated Systems, Vila do Conde, Portugal (2021). https://doi.org/10.1109/DCIS53048.2021.9666171

5. Al-Sarayreh, M., et. al.: Deep spectral-spatial features of snapshot HS images for red-meat classification. In: 2018 International Conference on Image and Vision Computing New Zealand (IVCNZ), pp. 1–6 (2018). https://doi.org/10.1109/IVCNZ.2018.8634783
6. Sancho, J., et al.: GoRG: towards a GPU-accelerated multiview hyperspectral depth estimation tool for medical applications. Sensors. **21**(12), 4091 (2021). https://doi.org/10.3390/s21124091
7. Wang, H., Peng, H., Chang, Y., Liang, D.: A survey of GPU-based acceleration techniques in MRI reconstructions. Quant Imaging Med. Surg. **8**(2), 196–208 (2018). https://doi.org/10.21037/qims.2018.03.07
8. J. Luitjens. Faster Parallel Reductions on Kepler. NVIDIA developer blog 2014; Online resource Accessed 21 Apr 11. https://developer.nvidia.com/blog/faster-parallel-reductions-kepler/
9. NVIDIA Jetson Linux Driver Package Software Features. Clock Frequency and Power Management. Release 32.6.1. NVIDIA documentation, 3 August 2021

Exploring Fully Convolutional Networks for the Segmentation of Hyperspectral Imaging Applied to Advanced Driver Assistance Systems

Jon Gutiérrez-Zaballa[1(✉)], Koldo Basterretxea[1], Javier Echanobe[2], M. Victoria Martínez[2], and Inés del Campo[2]

[1] Department of Electronics Technology, University of the Basque Country, 48013 Bilbao, Spain
j.gutierrez@ehu.eus

[2] Department of Electricity and Electronics, University of the Basque Country, 48940 Leioa, Spain

Abstract. Advanced Driver Assistance Systems (ADAS) are designed with the main purpose of increasing the safety and comfort of vehicle occupants. Most of current computer vision-based ADAS perform detection and tracking tasks quite successfully under regular conditions, but are not completely reliable, particularly under adverse weather and changing lighting conditions, neither in complex situations with many overlapping objects. In this work we explore the use of hyperspectral imaging (HSI) in ADAS on the assumption that the distinct near infrared (NIR) spectral reflectances of different materials can help to better separate the objects in a driving scene. In particular, this paper describes some experimental results of the application of fully convolutional networks (FCN) to the image segmentation of HSI for ADAS applications. More specifically, our aim is to investigate to what extent the spatial features codified by convolutional filters can be helpful to improve the performance of HSI segmentation systems. With that aim, we use the HSI-Drive v1.1 dataset, which provides a set of labelled images recorded in real driving conditions with a small-size snapshot NIR-HSI camera. Finally, we analyze the implementability of such a HSI segmentation system by prototyping the developed FCN model together with the necessary hyperspectral cube preprocessing stage and characterizing its performance on an MPSoC.

Keywords: hyperspectral imaging · scene understanding · fully convolutional networks · autonomous driving systems · system on chip

---

This work was partially supported by the Basque Government under grants PIBA-2018-1-0054, KK-2021/00111 and PRE 2021 1 0113 and by the Spanish Ministry of Science and Innovation under grant PID2020-115375RB-I00. We thank the University of the Basque Country for allocation of computational resources.

K. Desnos and S. Pertuz (Eds.): DASIP 2022, LNCS 13425, pp. 136–148, 2022.
https://doi.org/10.1007/978-3-031-12748-9_11

1 Introduction

Today, thanks to the availability of small-size, portable, snapshot hyperspectral cameras, it is possible to set-up HSI processing systems on moving platforms. The use of drones for precision agriculture and ecosystem monitoring is probably one of the most active and mature application domains [5]. The research into how hyperspectral information can be used to develop more capable and robust ADAS is, on the contrary, in its infancy [6,11,12]. HSI provides rich information about how materials reflect light of different wavelengths (spectral reflection), and this can be used to identify and classify surfaces and objects in an scene. Thus, with the application of appropriate information processing techniques, HSI can help to enhance the accuracy and robustness of current ADAS for object identification and tracking and, eventually, can be used for scene understanding, which is a step forward in the achievement of more capable and intelligent ADS (Autonomous Driving Systems).

HSI segmentation of real driving scenes is, however, challenging for a variety of reasons. First, the spectral reflectance signatures of the different objects, e.g. metallic white vehicles bodies and road marks, may be weakly separable. Second, extracting spatial features that could help segmenting items with similar spectral reflectances is difficult as a consequence of the enormous diversity of shapes, view angles and scales. Finally, it should be always kept in mind that developed segmentation algorithms need to be computed with very demanding latency requirements on resource constrained onboard processing platforms.

In this article we describe some results of a research that investigates how FCN can be applied to enhance the segmentation accuracy of images acquired in real driving scenarios with a small-size mosaic snapshot hyperspectral camera. We present a simple application example of scene understanding for the separation of the drivable (tarmac) and non-drivable areas (identifying sky and vegetation) in the acquired image sequences as well as for the recognition of road marks, which could be used to enhance automatic lane keeping and trajectory planning systems for ADS. Finally, we describe the rapid prototyping workflow used to develop a functional HSI segmentation processing system on a Xilinx Zynq UltraScale MPSoC, from algorithm exploration and model optimization to the final implementation.

2 Experimental Setup

When dealing with a semantic segmentation problem, it is of utmost importance to adapt the structure of the neural network to the unique characteristics of the dataset. Thus, once the suitability of the dataset has been verified, a hyperparameter tuning and optimization process should be carried out on the neural network.

2.1 The Dataset

As it is reported in [1], there are very few datasets of hyperspectral imagery for ADAS and ADS applications, one of which is precisely presented in [1], HSI

Drive. HSI Drive v1.1 contains 276 images of urban, road and highway scenarios in diverse weather (sunny, cloudy, rainy and foggy) and lightning (dawn, midday, sunset) conditions taken during Spring (121 images) and Summer (155 images).

The driving scenes have been recorded with a Photonfocus camera that includes an Imec 25-band VIS-NIR (535 nm–975 nm) sensor based on a CMOSIS CMV200 image wafer sensor. The global resolution is 1088×2048 pixels with $5\,\mu\text{m} \times 5\,\mu\text{m}$ size. However, as the spectral bands are extracted from a mosaic formed by 5×5 pixel window Fabri-Perot filters, the final resolution of the HSI cubes is $216 \times 409 \times 25$ [8]. This implies including a preprocessing stage in the processing pipeline that is addressed in Subsect. 4.1.

The original labelling separates the scenes into 10 classes taking into account the surface reflectances of the materials. Those classes are: Road, Road Marks, Vegetation, Painted Metal, Sky, Concrete/Stone/Brick, Pedestrian/Cyclist, Water, Unpainted Metal and Glass/Transparent Plastic. Furthermore, it has to be noted that the labelling of the dataset has followed a weak approach in order to provide the network with the most precise data. This means, for example, that pixels that are in the junction of two or more surfaces have been left out of the labelling process. However, these pixels do take part in the training process of a convolutional network. In fact, the training of convolutional neural networks with weakly labelled datasets is a line of research itself [10].

Spectral separability analysis measures the differences in the surface reflectance patterns of the materials that belong to different classes, which is an index of how well a semantic classifier could perform. One of the most common criteria in remote-sensing applications is JeffreysMatusita distance [4]. It ranges from 0 to 2 but does not have a linear interpretation as 0–1 values mean very poor separability, 1.0-1.9 values account for moderate separability and 1.9-2.0 values indicate good separability [1].

Table 1. JeffreysMatusita (JM) interclass distances.

	Road	Road M.	Veg.	P. Met.	Sky	Conc.	Ped.	Unp. Met.	Glass
Road		1.92	1.83	1.65	1.98	1.44	1.84	1.42	1.49
Road Marks	1.92		1.79	1.63	1.87	1.68	1.92	1.92	1.93
Vegetation	1.83	1.79		1.44	1.96	1.64	1.81	1.73	1.81
Painted Metal	1.65	1.63	1.44		1.91	1.48	1.70	1.35	1.48
Sky	1.98	1.87	1.96	1.91		1.97	1.98	1.98	1.80
Concrete	1.44	1.68	1.64	1.48	1.97		1.74	1.62	1.68
Pedestrian	1.84	1.92	1.81	1.70	1.98	1.74		1.79	1.66
Unpainted Metal	1.42	1.92	1.73	1.35	1.98	1.62	1.79		1.38
Glass	1.49	1.93	1.81	1.48	1.80	1.68	1.66	1.38	
Mean	**1.73**	**1.80**	**1.78**	**1.63**	**1.96**	**1.69**	**1.82**	**1.68**	**1.71**

Table 1 sums up the JM interclass distances of the ten classes. As it can be observed, the separability of some of the classes such as Road/Road Marks (1.92), Road/Sky (1.98) or Road Marks/Unpainted Metal (1.92) is promisingly high while classes like Road/Concrete (1.44), Road Marks/Painted Metal (1.63) and Painted Metal/Unpainted Metal (1.35) show low separability indexes.

2.2 FCNs for HSI Image Segmentation

The neural network selected to perform semantic segmentation is a typical FCN known as U-Net [9] which was originally intended for biological image segmentation but has been widely used for other segmentation tasks, such as, precision agriculture [10] and aerial city recognition [3]. The idea of using a FCN is to combine the intrinsic spectral characteristics of the different classes with the spatial relationships that should be extracted by the convolution operations.

We have adapted the original architecture of the U-Net [9] to the unique characteristics of the dataset to achieve the best trade-off between segmentation performance and computational complexity. With this aim we have performed a grid search of the optimum combination of model hyperparameters by evaluating the segmentation accuracy on a subset of 45 images selected from all possible environment/weather conditions.

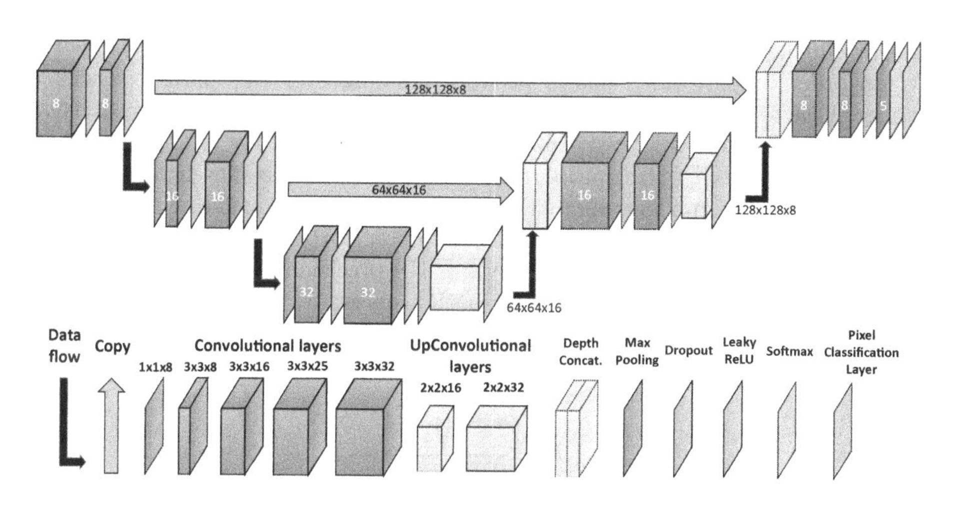

Fig. 1. Architecture of the modified U-Net.

The set of analyzed hyperparameters included: the size of the input image patches, the overlapping between patches, the encoder depth and the number of filters in the first convolutional block. In order to avoid an unaffordable optimization time, we have consulted the typical values of the hyperparameters to be optimized in the literature [10]. This way, a specific range has been set for each hyperparameter. Specifically, the value of the encoder depth has been varied between 2 and 4, the number of initial filters between 8 and 32 (in powers of 2), while for the patch size the values 64 and 128 have been evaluated.

The values of the hyperparameters that have output the three best results are: 2-3-4 for the encoder/decoder depth, 8-16-16 for the initial number of filters and 128-128-128 for the side of the square patch. According to the accuracy/complexity trade-off criterion we have selected the 2/8/128 set (Fig. 1 shows the final architecture of the network). As a consequence of the size of the patch and to benefit from the effect of overlapping, it has been decided to divide the input test images in 18 (3×6) patches.

3 Segmentation Results

The first experiment focuses on segmenting 3 classes: Road, Road Marks and No Drivable (the remaining classes). The proposed low-complexity segmentation system would be aimed at a possible final system for the discrimination of drivable and non-drivable zones, together with a lane-keeping aid.

In a second experiment we have added two additional classes to the model training; Vegetation and Sky. These two categories have been selected due to their satisfactory spectral separability indexes (see Table 1). The exploration of more complex segmentation models including all classes in the dataset has also been performed but obtained results are irregular and not concluding, and will require further investigation.

In order to perform a neural network training over this dataset, the 276 images have to be divided into training, validation and test subsets. This division has been performed as follows: 162 images for training, 57 for validation and 57 for testing, preserving class proportionality in all the three subsets.

The chosen metrics to evaluate the segmentation ability of the neural network are accuracy, precision and intersection over union (IoU). As Eqs. 1, 2 and 3 show, accuracy accounts for the false negatives (FN), precision takes into account the false positives (FP) and IoU combines both aspects:

$$A_i = \frac{TP_i}{TP_i + FN_i} \tag{1}$$

$$P_i = \frac{TP_i}{TP_i + FP_i} \tag{2}$$

$$IoU_i = \frac{TP_i}{TP_i + FN_i + FP_i} \tag{3}$$

where i is the class index such that, for example, FN_i accounts for the pixels that have been predicted as not belonging to class i, but are actually part of class i.

As a consequence of the dataset being heavily imbalanced (the number of pixels in the test dataset is: Road 2,067,379; Road M. 99,426; Veget. 820,804; Sky 163,127 and Other 363,345) it is useful not to only represent the global metrics but also the mean values and, more specifically, the weighted scores. In order to do that, some weighting factors, which are related to the inverse of the frequency of the classes in the dataset, have been previously computed.

3.1 U-Net

Table 2 collects the performance, in accordance with the above mentioned metrics, of the modified U-Net and also the segmentation metrics after the overlapped patches have been joined to reconstruct the images to their original resolution. The comparison depicts that the use of overlapping patches improves the segmentation, specially the precision, compared to the case in which the patches do not overlap. This is because neural networks tend to fail to predict the pixels of the patch contours because they lack surrounding information.

Table 2. Performance of the modified U-Net (patches and overlapping patches) and ANN on the 3-classes (up) and 5-classes (down) test datasets.

	U-Net						ANN		
	Patches (128x128x25)			Rebuilt images from overlapping patches			Pixels (1x1x25)		
	Accuracy	Precision	IoU	Accuracy	Precision	IoU	Accuracy	Precision	IoU
Road	97.90	95.66	93.74	98.54	94.56	93.25	85.10	92.51	79.62
Road Marks	90.25	73.11	67.75	87.89	77.22	69.80	68.10	21.90	19.86
No Drivable	91.07	97.16	88.71	91.20	98.57	90.01	86.46	89.32	78.42
Overall	95.37	95.55	91.31	95.42	95.44	91.50	85.14	89.32	77.46
Mean	93.07	88.64	83.40	92.54	90.12	84.35	79.89	67.93	59.30
Weighted	90.60	75.71	70.27	88.54	79.43	72.60	70.04	29.36	26.27
Road	92.61	99.05	91.36	93.28	99.00	92.41	73.29	94.30	70.18
Road Marks	80.93	75.39	64.02	78.32	79.11	64.90	68.74	17.50	16.21
Vegetation	94.98	94.63	90.12	95.74	95.80	91.88	93.84	91.47	86.29
Sky	97.86	93.09	91.23	97.49	93.39	91.20	91.04	74.53	69.44
Other	84.97	62.71	56.45	84.83	64.59	57.90	56.18	42.94	32.17
Overall	91.79	93.29	86.47	92.75	93.80	87.66	77.02	85.24	68.45
Mean	90.18	84.98	78.64	89.93	86.38	79.66	76.62	64.15	54.86
Weighted	88.64	82.14	75.27	87.40	84.15	75.93	75.28	44.00	39.55

Analyzing the numerical results of the reconstructed images, it can be seen that all the classes have a great IoU with the exception of the class Road Marks which suffers from a low precision value; in the first experiment, in particular, for every 100 TPs of its class there are 37 FPs. However, as Road Marks is the minority class, this value does not affect the overall result as Fig. 2 shows.

Figure 2 also depicts how the proposed FCN perfectly segments a typical driving scene (second column) for the 3-class experiment (second and third rows) while it fails to correctly identify some pixels in challenging images such as those where there are objects casting their shadows on the road (first column, especially in the background) or overlapping objects (third column, where the left side is populated with objects of different materials).

Table 2 also confirms the good segmentation of the 5-class experiment. The global result can be seen in Fig. 2 where, once again, the overall segmentation would be very useful in a lane-keeping system and would also allow the driver to

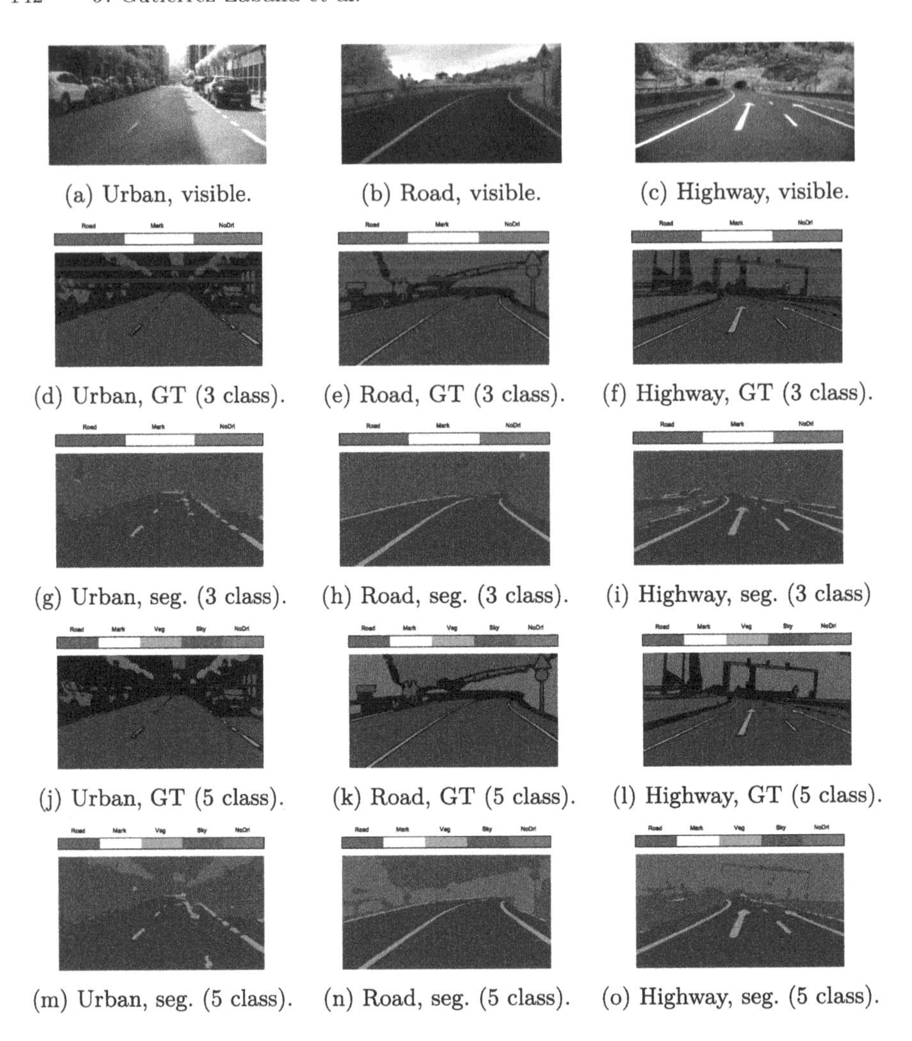

(a) Urban, visible. (b) Road, visible. (c) Highway, visible.

(d) Urban, GT (3 class). (e) Road, GT (3 class). (f) Highway, GT (3 class).

(g) Urban, seg. (3 class). (h) Road, seg. (3 class). (i) Highway, seg. (3 class)

(j) Urban, GT (5 class). (k) Road, GT (5 class). (l) Highway, GT (5 class).

(m) Urban, seg. (5 class). (n) Road, seg. (5 class). (o) Highway, seg. (5 class).

Fig. 2. Comparison among the visible (first row), 3-class ground truth (second row), 3-class U-Net segmentation (third row), 5-class ground truth (fourth row) and 5-class U-Net segmentation (fifth row) images of three different scenarios: urban (first column), road (second column) and highway (third column).

have more information about the surroundings. For instance, it can be observed that the system is now able to identify the presence of some objects in the no-drivable sections of the images such as traffic signals, pedestrians and guardrails.

3.2 A Comparison with Baseline Spectral Classifiers

From the above described results it can be concluded that the contribution of the spatial information provided by the convolution filters is, indeed, relevant to overcome the limitations inherent to the spectral separability of the different objects that can be present in real driving scenes. In order to get a more precise picture of this contribution we have compared the obtained results to those achieved with a baseline purely spectral classifier based on a three-hidden-layer feedforward ANN. The exploration and optimization process carried out has concluded with 25-25-100-100-3 being the best structure for the network. Table 2 gathers the metrics for the two experiments of the three-hidden-layer neural network.

The results achieved by the ANN are, all in all, far from the ones associated to the U-Net (specially in terms of the precision of the minority class), so it can be confirmed that the joint use of spatial and spectral information is beneficial for the segmentation ability of a neural network.

The difference in performance can also be explained in terms of model size and computational complexity (MACs, Multiply and Accumulate operations): while the ANN has only 13,855 parameters and performs 1,203,687,000 MACs (13,625 MACs per pixel) during inference, the U-Net has 31,725 parameters (320 of them are non-trainable) and needs 2,543,321,088 MACs (141,295,616 MACs per patch) to produce an output. The difference between the MACs ratio (2.11x) and the parameters ratio (2.3x) affects the time needed to make the forward pass, which will be assessed in the next section.

It is worth mentioning a relevant outcome relative to the significance of the information provided by the spectral bands of the sensor. Despite the high correlation observed between spectral bands in the dataset images, we have verified that the reduction of the spectral bands used in the training of spectral classifiers strongly conditions the achievable accuracy of the segmentation.

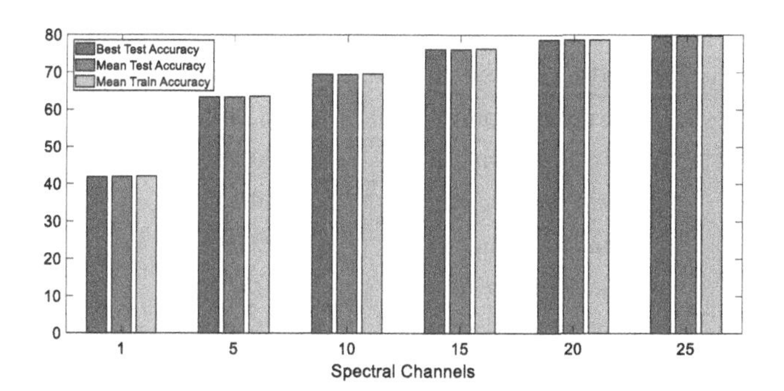

Fig. 3. Overall accuracy (%) as a function of the number of spectral channels.

As an example, Fig. 3 shows the overall accuracy results of a reference ELM (Extreme Learning Machine) classifier for a different number of spectral bands (ranging from just 1 to all the 25 bands). It can be noted that accuracy indexes vary almost 40%. On the contrary, we have observed that reducing the spectral bands in the training of the U-Net has not such a strong effect and the spatial information can compensate to a great extent the reduction in the spectral information provided as input. In particular, we have observed that when using just one spectral band (a gray level image, actually) the segmentation performance has only degraded, in terms of overall accuracy, by 0.75% when that band is the first principal component of a PCA (Principal Component Analysis) and by 1.75% when that band is just 1 of the 25 bands.

4 Workflow for Rapid Prototyping

As the implementation of this kind of neural networks in SoCs is a challenging and time consuming process, we have decided to explore the use of high-level automatic code generation tools to achieve a rapid prototyping of the system.

4.1 Image Preprocessing

Raw images acquired from mosaic snapshot cameras need to undergo a preprocessing pipeline in order to be converted into hyperspectral cubes. This process, which starts with raw image cropping and finishes with band normalization, needs to be taken into account when characterizing the throughput of the whole segmentation system. The rest of the steps are reflectance correction, partial demosaicing (original resolution is not restored), band alignment and spatial filtering. This processing has been codified in C language and compiled to be executed as an embedded Linux application in the microprocessor of the MPSoC as part of the HW/SW codesign for the implementation of the system.

Table 3. Mean execution time of the image-preprocessing Linux application.

Step Name	Execution time (ms)
Image cropping	4.36
Reflectance correction	68.49
Partial demosaicing	30.15
Band alignment	22.41
Spatial filtering	202.65
Band normalization	26.06
Total	353.97

Table 3 shows the mean latency over 1000 iterations of the preprocessing pipeline running on the Cortex A-53 Quadcore processor (which has NEON SIMD extensions mandatory per core) in the Zynq MPSoC which reaches a reasonable value of 353.97 ms.

4.2 Neural Network Deployment

The design, training, validation and test of the FCN has been performed using MATLAB's Deep Network Designer. For the segmentation system prototyping process on the MPSoC we have used Vitis AI, a development platform for AI inference on Xilinx hardware platforms. Since there is not a direct procedure for the implementation of MATLAB-generated deep models we had to first export the neural network to open neural network exchange (ONNX) representation and then import it to Keras, our chosen framework, via *onnx2keras*, an ONNX to Keras neural network converter [7]. The next steps are the freezing and quantizating processes of the neural network that are necessary due to the fact that Vitis AI favours integer computing.

The workflow continues by creating a Tensorflow inference graph from the Keras model and by removing the unnecessary information from training and saving only the required elements to compute the requested outputs. As it is known, inference is computationally expensive and, in order to reach the high-throughput and low-latency requirements of ADAS applications, a high memory bandwidth is required. Vitis AI Quantizer exploits quantization and VAI Optimizer applies channel pruning techniques to meet those issues.

According to [13], by converting the 32-bit floating point weights and activations to 8-bit integer format, Vitis AI quantizer can reduce computing complexity without losing prediction accuracy and, as the fixed-point network model requires less memory bandwidth, a faster speed is provided.

Finally, the product of the quantization is loaded at runtime in the system composed by the ARM CPU and the DPU accelerator in the MPSoC by a Python VART API. Although the quantization output of the 3-class experiment presents no appreciable IoU degradation, it has to be stated that the 5-class quantized model experiences a noticeable loss of performance on some images, an issue to be addressed in the future by Quantization Aware Training or Finetuning [13].

Figure 4 shows the output of the deployed model and confirms its good overall performance. In fact, the similarity between quantized and unquantized results is 97.82%, 98.16% and 98.66%, respectively.

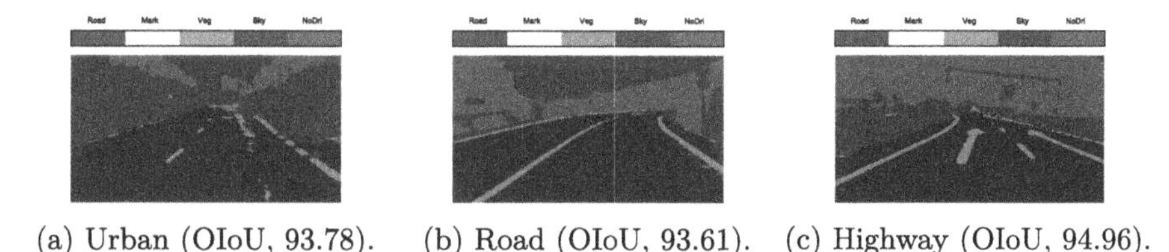

(a) Urban (OIoU, 93.78). (b) Road (OIoU, 93.61). (c) Highway (OIoU, 94.96).

Fig. 4. Segmented images produced by the deployed model on the MPSoC.

In terms of throughput, the inference of the U-Net deployed in the Zynq UltraScale+ MPSoC reaches a rate of 487.91 FPS when 2 DPUCZDX8G cores (B4096 architecture) are involved, that is, an image (18 patches) is segmented every 36.89 ms (27 FPS). In [2] the FPS of some of the state-of-the-art neural networks for segmentation are evaluated when run on a GTX 1080 Ti and range from 5 to 23 FPS. However, if we add to it the time employed during the preprocessing (Table 3), the throughput decreases to 2.55 FPS which can be considered a good starting point but will need to be improved by accelerating by hardware some of the preprocessing steps or changing the band extraction approach.

We have also benchmarked three different device types and evaluated their performance in terms of throughput with comparative purposes: an Intel-Xeon E5 1620 v3 Quadcore (CPU) and two embedded platforms such as a Jetson Nano (GPU) and a Xilinx ZCU104 (FPGA).

Table 4. Throughput of the U-Net and ANN on different device types

	U-Net (18x128x128x25)				ANN (216x409x25)			
Device type \ Timing (FPS)	Mean	Median	Max	Min	Mean	Median	Max	Min
CPU (Intel Xeon), FP32	1.49	1.51	1.62	0.19	2.49	2.51	3.29	1.28
GPU (Jetson Nano), FP16/FP32	10.70	10.76	11.80	3.07	17.74	17.90	19.50	6.12
FPGA (Zynq Ultrascale+), INT8	27.11	27.70	29.39	19.52	-	-	-	-

Table 4 displays the mean, median, maximum and minimum FPS values of 1000 executions of the segmentation of one image with both neural networks on each device. It shows, on the one hand, how the ANN is 1.66x faster compared to the U-Net although we have to take into account that U-Net performs only 2.11x MACs to get a much better segmentation output. On the other hand, it also confirms the benefits of using a dedicated custom hardware such as the DPU included in the PL part of the MPSoC which outperforms the U-Net CPU and GPU approaches (18.2x and 2.5x FPS ratios respectively) and even exceeds the ANN GPU alternative (1.5x FPS ratio).

5 Conclusions

The incorporation of richer spectral information through HSI improves the segmentation results of purely spectral models. Besides, it is confirmed that the use of spatial information via convolution operations outperforms purely spectral models, even when dealing with images as intricate and heterogeneous as those that must be processed in real driving scenarios. However, the contribution of the spectral information in spectro-spatial convolutional models needs to be further investigated since our experiments reveal that the spectral information is being overshadowed by the spatial information in the training process of the FCN segmentation. We expect that the more effective incorporation of the spectral information to the AI models should improve segmentation performance in tricky situations such as when there are areas with shadows, there is degradation in the materials to be segmented, there are surfaces with very high reflectance in conditions of extreme lighting or there are multiple overlapping objects.

The improvement of the segmentation performance involves investigating further modifications to the proposed U-Net such as using 3D convolutions or applying multiscale convolution techniques to extract spatial features at different scales. The use of different image preprocessing techniques (modifying the partial demosaicing step, for example) and the addition of a postprocessing stage (not to label pixels with uncertain prediction, for instance) will also be assessed, but their applicability will always be subject to the demanding throughput requirements of ADAS/ADS.

In turn, a workflow which combines the use of MATLAB's (Deep Learning and Image Processing toolboxes) and Xilinx's tools (AI development and deployment environments) for the rapid prototyping of AI applications has been set. This will allow us to perform agile deployment characterization of future models so as to rapidly evaluate their suitability to ADAS applications.

Finally, we have verified that the combination of an encoder-decoder FCN and the prototyping platform (Zynq Ultrascale MPSoC) allows us to perform the preprocessing of the cubes and the segmentation of the images in a reasonable time for this kind of applications outperforming CPU-only and GPU approaches. However, there are multiple pathways for future research, either in terms of the acceleration of the cube generation (hardware acceleration of the median filtering) or the optimization of the U-Net. At the same time, a deeper investigation is required regarding the data transfer between memory and the different components of the SoC to try to improve the throughput and reduce the cost and consumption.

References

1. Basterretxea, K., Martínez, V., Echanobe, J., Gutiérrez-Zaballa, J., Del Campo, I.: Hsi-drive: a dataset for the research of hyperspectral image processing applied to autonomous driving systems. In: 2021 IEEE Intelligent Vehicles Symposium (IV), pp. 866–873 (2021). https://doi.org/10.1109/IV48863.2021.9575298
2. Courdier, E., Fleuret, F.: Real-time segmentation networks should be latency aware. In: Proceedings of the Asian Conference on Computer Vision (2020)
3. Cui, X., Zheng, K., Gao, L., Zhang, B., Yang, D., Ren, J.: Multiscale spatial-spectral convolutional network with image-based framework for hyperspectral imagery classification. Remote Sens. **11**(19), 2220 (2019)
4. Forestier, G., Inglada, J., Wemmert, C., Gançarski, P.: Comparison of optical sensors discrimination ability using spectral libraries. Int. J. Remote Sens. **34**(7), 2327–2349 (2013)
5. Govender, M., Chetty, K., Bulcock, H.: A review of hyperspectral remote sensing and its application in vegetation and water resource studies. Water Sa **33**(2), 145–151 (2007)
6. Huang, Y., Huang, E., Chen, L., You, S., Fu, Y., Shen, Q.: Hyperspectral image semantic segmentation in cityscapes. arXiv preprint arXiv:2012.10122 (2020)
7. Malivenko, G.: onnx2keras 0.0.24. https://pypi.org/project/onnx2keras/ (2021)

8. Photonfocus: MV1-D2048x1088-HS02-96-G2. https://www.photonfocus.com/products/camerafinder/camera/mv1-d2048x1088-hs02-96-g2
9. Ronneberger, O., Fischer, P., Brox, T.: U-Net: convolutional networks for biomedical image segmentation. In: Navab, N., Hornegger, J., Wells, W.M., Frangi, A.F. (eds.) MICCAI 2015. LNCS, vol. 9351, pp. 234–241. Springer, Cham (2015). https://doi.org/10.1007/978-3-319-24574-4_28
10. Wang, S., Chen, W., Xie, S.M., Azzari, G., Lobell, D.B.: Weakly supervised deep learning for segmentation of remote sensing imagery. Remote Sens. **12**(2), 207 (2020)
11. Winkens, C., Sattler, F., Adams, V., Paulus, D.: Hyko: a spectral dataset for scene understanding. In: Proceedings of the IEEE International Conference on Computer Vision Workshops, pp. 254–261 (2017)
12. Winkens, C., Sattler, F., Paulus, D.: Hyperspectral terrain classification for ground vehicles. In: VISIGRAPP (5: VISAPP), pp. 417–424 (2017)
13. Xilinx: Quantizing the model. https://www.xilinx.com/html_docs/vitis_ai/1_4/quantize.html#uim1570695919827 (2021)

An Adaptable Cognitive Microcontroller Node for Fitness Activity Recognition

Matteo Antonio Scrugli[1(✉)], Bojan Blažica[2], and Paolo Meloni[1]

[1] Department of Electrical and Electronic Engineering (DIEE), University of Cagliari, Cagliari, Italy
matteo.scrugli@unica.it

[2] Computer Systems Department, Jožef Stefan Institute, Ljubljana, Slovenia

Abstract. The new generation of wireless technologies, fitness trackers, and devices with embedded sensors can have a big impact on healthcare systems and quality of life. Among the most crucial aspects to consider in these devices are the accuracy of the data produced and power consumption. Many of the events that can be monitored, while apparently simple, may not be easily detectable and recognizable by devices equipped with embedded sensors, especially on devices with low computing capabilities. It is well known that deep learning reduces the study of features that contribute to the recognition of the different target classes. In this work, we present a portable and battery-powered microcontroller-based device applicable to a wobble board. Wobble boards are low-cost equipment that can be used for sensorimotor training to avoid ankle injuries or as part of the rehabilitation process after an injury. The exercise recognition process was implemented through the use of cognitive techniques based on deep learning. To reduce power consumption, we add an adaptivity layer that dynamically manages the device's hardware and software configuration to adapt it to the required operating mode at runtime. Our experimental results show that adjusting the node configuration to the workload at runtime can save up to 60% of the power consumed. On a custom dataset, our optimized and quantized neural network achieves an accuracy value greater than 97% for detecting some specific physical exercises on a wobble board.

Keywords: Adaptive system · Fitness activity tracking · Sensorimotor training · Low power electronics · Neural network · Remote sensing · Runtime

1 Introduction

The guidelines of the World Health Organization (WHO) in 2010 document, excluding special cases, an average adult should engage in physical activity of moderate intensity for at least 150 min per week and 75 min per week at high intensity [1]. Tracking and encouraging good levels of physical activity can improve people's health [2]. Fitness tracker devices have had a rapid development in recent

K. Desnos and S. Pertuz (Eds.): DASIP 2022, LNCS 13425, pp. 149–161, 2022.
https://doi.org/10.1007/978-3-031-12748-9_12

years, due to their ease of use, accuracy, and portability. Events in a trackable signal, although seemingly simple, can be difficult to identify and recognise within the data stream. Deep learning is well known for reducing the study of features that contribute to the recognition of different target classes and greatly increase the classification accuracy, but in order to be used in low power devices a careful software optimization is necessary.

In this work, we present a portable and battery-powered microcontroller-based device applicable to a wobble board. Wobble boards are inexpensive and easy-to-use tools to avoid ankle injuries or as part of the recovery process after an injury (Fig. 1). The exercise recognition process was implemented through the use of cognitive techniques based on deep learning.

Fig. 1. Wobble board used to validate our approach.

To manage the hardware/software configuration we have implemented a component called ADAM (ADAptive runtime Manager), able to optimize device power consumption and performance. ADAM creates and manages a network of processes that communicate with one another via FIFOs. The morphology of the process network varies depending on the operating mode (OM) in execution. ADAM can be triggered by external environment re-configuration messages or by specific workload-related variables in the sampled streams. When triggered, ADAM alters the morphology of the process network by turning on or off processes and rearranging the inter-process FIFOs. Furthermore, depending on the new configuration, it modifies the platform's hardware configuration, adjusting power-related settings such as clock frequency, supply voltage, and peripheral gating.

2 Related Work

Local processing is frequently used only for implementing simple checks on raw data and/or marshaling tasks for wrapping sensed data inside standard communication protocols, and the edge-computing paradigm is only marginally exploited [3–5]. More complex and accurate algorithms, such as those based on artificial intelligence or deep learning, must be targeted in order to properly use cognitive computing at the edge. Their effectiveness on high-performance computing platforms has been widely demonstrated. Nonetheless, how to map state-of-the-art cognitive computing on resource-constrained platforms remains an open question. To identify specific events in sensed data, an increasing number of approaches based on machine learning and artificial intelligence are being developed. In [6] and [7] ANN (artificial neural networks) are used by the authors to detect specific conditions in the proposed data. In [7], an ANN is used to determine the patient's emotional state (happiness or sadness). In our previous work [8], energy/power efficiency is improved using ADAM component, using near-sensor processing to save data transfers, and dynamically adapting application setup and system frequency to the OM requested by an external user and to data-dependent workload.

In our use case, a CNN (Convolutional Neural Network) is used to identify and recognize simple physical exercises performed on a wobble board. In [9,10], the authors propose the use of a wobble board in creative way, to entice people to its use, as it is very important in ankle rehabilitation. In a review of the concept of patient motivation [11], the authors describe how motivation has been considered in relation to rehabilitation associated with strokes, fractures, rheumatic disease, aging, and cardiac and neurological issues. The limited motivation on the part of some individuals may, at least in part, be ascribed to the tedious nature of the ankle exercises and the inability to monitor one's improvement throughout the course of the training process [12].

A similar approach is considered in our work, the implemented system detects and identifies some simple sensorimotor exercises performed on the wobble board, giving a percentage of correct execution at the end of the exercise. As far as we know, our system is the only one that applies state-of-the-art deep learning-based techniques to recognize some specific movements on a wobble board, managing hardware and software dynamically in order to minimize power consumption.

3 Wobble Board and Node Architecture

We used a wobble board capable of 360° rotation (Fig. 1), the sensory node is fixed in the upper-middle part. Since the device is battery powered and communication is via a Bluetooth Low Energy module, no cable is used to interface with the sensor node. We chose STMicroelectronics SensorTile microcontroller-based platform, which is equipped with an ARM Cortex-M4 32-bit low-power microcontroller. It takes advantage of the LSM303AGR accelerometer sensor integrated into the Sensortile, only the two axes X and Y parallel to the floor are taken into account and, for each sensor, it is necessary to make a calibration that takes into account the offset on the acquired data. It was chosen to run FreeRTOS on the node, to have more control over the running tasks due to its ability to create a thread-level abstraction.

3.1 Application Model

We chose a process network-based application structure. Tasks are modeled as separate processes that communicate with one another via FIFO structures. Using a software pipeline, processes can potentially be executed in parallel, improving performance. When the topology of the network processes changes, a change of OM occurs.

We identified four topologies of processes that can be combined in different ways:

- ***Get data task***: take data from the sensing hardware.
- ***Process task:*** it's possible to have multiple tasks of this type, representing multiple stages of in-place data analysis algorithm.
- ***Threshold task:*** filters data depending on the results of the analysis.
- ***Send task:*** is the task in charge of outwards communication to the gateway.

3.2 Adaptivity Support: The ADAptive Runtime Manager

A task within the process network was dedicated to the management of the platform's dynamic hardware and software reconfiguration. We have implemented such reconfiguration in a software agent called ADAptive runtime Manager (ADAM). ADAM can be activated on a regular basis by using an internal timer, it monitors the system's status, such as changes in workload. ADAM can react to such input by changing the platform settings, performing various operations such as enabling or disabling individual tasks of the sensor task chain or the entire chain; deciding whether to put the microcontroller in sleep mode or not; setting the operating frequency; and rerouting the data-flow managed by the FIFOs based on the active tasks.

4 Designing the Application

In this work, a system is implemented that is able to recognize typical movements in exercises that involve the use of a conventional wobble board (described in Sect. 3) or, more simply, the wireless transmission of raw data acquired from the sensor. The application model chosen for this use case provides two possible levels of processing able to evaluate the nature of the movement. The OMs chosen for the selected use case are shown in Fig. 2 and described below.

4.1 Operating Mode: Raw Data

This is the simplest OM, using only two tasks. It is possible to acquire data from the sensor and send it via Bluetooth, with a sampling frequency 100 Hz. In order to reduce the power consumption related to the transmission, it was decided to encapsulate four samples taken from the sensor in a single low energy Bluetooth packet. The Bluetooth packet has a size of 20 bytes, 4 bytes the timestamp, and four pairs of data taken from the sensor at different instants of time, the data pair is formed by the values relative to the accelerometer's X and Y axis, each with a size of 2 bytes.

4.2 Operating Mode: Basic Balance

This OM enables the first level of processing. The sampling rate is lowered to 100/7 Hz, which is more than sufficient to perform the analysis in this OM. However, given the storing signal frames and the size of the neural network (highly dependent on the input size), higher sampling rate values lead to memory footprint issues for the selected reference platform. A simple algorithm calculates how much, in percentage, the wobble board is in a balanced position. The extreme cases, the analysis returns a value of 100% if the board remains horizontal within a certain tolerance and 0% when the board remains in constant contact with the ground. The result of the analysis is transmitted every second, this leads to a significant energy saving due to the decrease of information that has to be sent via Bluetooth, which is no longer used to transmit raw data.

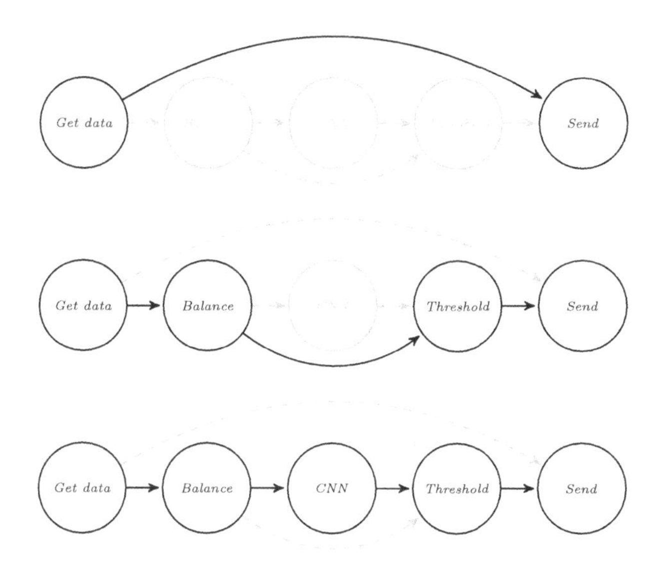

Fig. 2. Application model. Top *raw OM*, middle *balance OM*, bottom *CNN OM.*

4.3 Operating Mode: CNN

Some exercises were selected which were not too complicated to be recognized by deep learning techniques. Also in this case, a frequency of 100/7 Hz is ideal to obtain good results with the neural network and have a not excessive workload. The raw data related to the two X and Y axes of the accelerometer are used as two different input features as input to the neural network. The *balance* task remains active so that if a total stop of the table is detected, no CNN is executed. Again, the result of the analysis is transmitted every second. Some typical exercises recommended by Anders Heckmann [12] are those shown in Fig. 3.

The correct execution of the exercise involves:

- **Basic stance balance (Fig. 3.a):** Stand on the board with the edges of your feet on the outer edges of the board. Maintain a neutral spine and keep your torso upright. Balance on the board by shifting your weight to prevent any of the board's edges from touching the floor. The goal is to maintain the balance for 60 s.
- **Side tilt (Fig. 3.b):** Stand on the wobble board with your feet on the outer edges. Stand upright in a neutral spine position. Tilt the board from left to right by transferring your weight from your left leg to your right leg. Moving in a slow and controlled manner, keeping an upright torso and tight core. The duration of the exercise is 60 s.

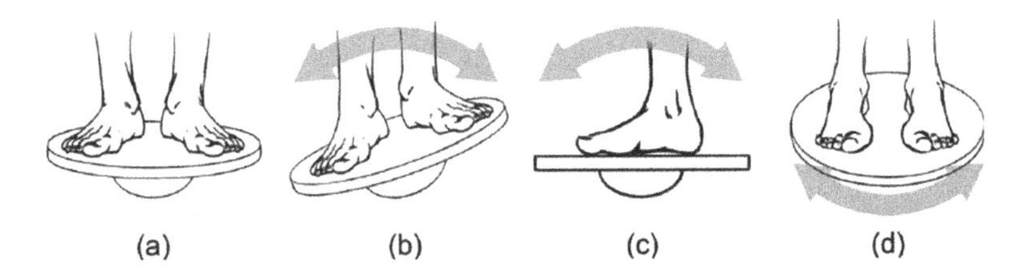

Fig. 3. The four common wobble board exercises recommended by physiotherapist Anders Heckmann [12]. (a) Balance while keeping as steady as possible. (b) Move the board from side to side. (c) Move the board back and forth. (d) Clockwise and counterclockwise circular movement. The Figure was extracted from [9]

- **Forward/backward tilt (Fig. 3.c):** Stand on the wobble board with your feet on the outer edges. Stand upright in a neutral spine position. Tilt the board to the front to touch the floor. Tilt it back onto the heels to touch the floor behind you. Continue tilting forward and back in a slow, steady, controlled motion for 60 s.
- **Two leg tilts (Fig. 3.d):** Stand on the wobble board with your feet on the outer edges. Stand upright in a neutral spine position. In a combination of the two previous exercises, you will roll the board in a 360-degree motion. Begin by tilting the board to the left. When the board touches the ground on the left, transfer your weight to the front to touch the floor. Now transfer your weight to touch the floor to the right side. Complete the revolution by tilting the board to the floor behind you. Keep your body centralized throughout. You may need to balance with your arms as you get used to the movement. Reverse the motion to move in the other direction. Continue for 60 s.
- **Other:** there is a fifth class that represents everything that is not foreseen by the previous exercises, for example the fall from the table or the absolute absence of movement.

4.4 Neural Network Design

We used a training procedure that included a static quantization[1] step, the source code is available in our public repository[2]. This process converts floating-point weights and activations to integers, allowing the CNN to be implemented using the CMSIS-NN optimized function library, which expects inputs with 8-bit precision. We have chosen to force the value of the bias to zero, while for the conversion of the weights we have inserted *MinMax* observers[3], who have the task of studying the outputs of each layer. Evaluating the distribution of the output values of each layer allows the observer to establish a value of *scale* and *zero-point* in order not to saturate these values using a quantized network. The CMSIS-NN library's functions for implementing convolution and fully connected layers include output shifting operations for applying the Scale factor to the outputs, with scaling values ranging from -128 to 127. In PyTorch, however, the quantization procedure requires a Scale value that is not always a power of two. As a result, we modified the CMSIS functions slightly to support arbitrary *scale* values. This change resulted in a minor increase in inference execution time. After testing inference with and without this modification, we calculated an increase in execution time of 2.87%.

We used a design space exploration process to compare tens of neural network topologies in terms of accuracy achieved after training and computing workload associated with executing the inference task on SensorTile. Figure 4 shows the selected convolutional network and the five selected output classes. All exercises are one minute in length, each movement performed during the four different exercises has a different duration. Generally, the longest exercise is the two leg tilts. CNN does not analyze the exercise for its entire duration (60 s) at once, the signal is divided into windows of 15 s duration, a good compromise between temporal precision and distinction between the movements to be evaluated. The maximum number of epochs was set to 30 and the Early Stopping (ES) algorithm was chosen to avoid overfitting effects. This algorithm terminates the training phase if it detects an increase in the loss value [13]; the loss is evaluated every epoch, and a Patience value of 5 is selected, implying that the training terminates only if a loss increment is detected for 5 consecutive epochs. Table 1 summarizes the parameters chosen for training, while Table 2, shows the dimensions of the various layers chosen.

[1] https://pytorch.org/tutorials/advanced/static_quantization_tutorial.html.
[2] https://github.com/matteoscrugli/deepwobbleboard.
[3] https://pytorch.org/docs/stable/_modules/torch/quantization/observer.html.

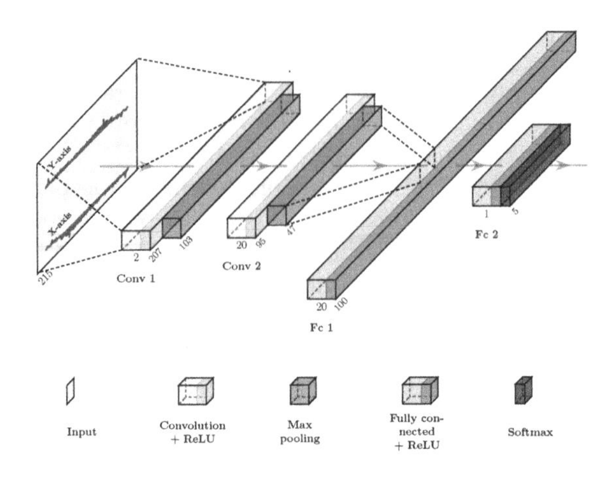

(B) Basic stance balance	**(FB)** Forward/backward tilt
(S) Side tilt	**(R)** Two leg tilts
(G) Other	

Fig. 4. CNN structure and classes description.

Table 1. Hyperparameters used during the training phase.

Hyperparameter	*Value*	*Hyperparameter*	*Value*
Epochs	30	Optimizer	Adadelta
Batch size	32	Learning rate	1.0
Loss criterion	Cross Entropy	Rho	0.9
ES patience	5	ES evaluation	Every epoch

Table 2. Model parameters.

Layer	*Input dimension*	*Output dimension*	*Input features*	*Output features*	*Kernel size*
Convolutional	215	207	2	20	9
Max pooling	207	103	20	20	2
Convolutional	103	95	20	20	9
Max pooling	95	47	20	20	2
Fully connected	940	100	–	–	–
Fully connected	100	5	–	–	–

4.5 Data Augmentation and Generalization

Unfortunately, no datasets already used in the literature containing the selected exercises were found. Therefore, a dataset was created with 12 one-minute recordings for each type of exercise (including class "other"). A random split of the dataset was chosen in order to use 80% of the data for the training set and 20% for the validation set. Operations such as translation, rotation and time dilation of the signal in each direction can often greatly improve generalization [13]. An exploration was made concerning the parameters used for all the agumentation techniques selected, In this paper, the exploration phase of these parameters will not be elaborated. Parameters were selected that led to better results during test set inference and thus actually helped to improve the training phase of the neural network.

We planned to expand our dataset in order to increase the generality of the dataset and without necessarily resorting to the use of data augmentation techniques.

In more detail, augmentation techniques are (also summarized in Table 3):

- **Translation:** During training, the window to the entire signal is shifted by 0.25 s per frame. For example, a 60-second recording with a translation of 0.25 generates a number of frames of $(60 - 15)/0.25 + 1 = 181$.
- **Rotation:** A rotation transformation was applied to the X and Y axes of the sensor data, in our case we chose two rotations of $\angle{-4}$ and $\angle 4°$. For each record, two more are then generated.
- **Dilation:** The sampling frequency of the signals in the dataset 100 Hz, but the neural network is trained with 100/7 Hz signals. The size of the input signal is therefore equal to $\lfloor(15 \times 100 + 7 - 1)/7\rfloor = 215$. Time dilation can be obtained by increasing or decreasing the downsampling while keeping the input size to the neural network constant, in this case, two additional downsampling values of 6 and 8 were chosen. For each recording, two more are generated with different time dilations.

Table 3. Augmentation parameters.

Parameter	*Value*
Traslation, temporal distance between frames	0.25 s
Rotation, X and Y axis rotation	$\angle{-4}$, $\angle 0$, $\angle 4$
Time dilation, downsampling	6, 7, 8

5 Experimental Results

In this section, we will show the results obtained after the neural network training and we will make a detailed analysis of the power consumption for each OM.

5.1 Neural Network Accuracy

After the training phase, an accuracy of 97.652% was measured on the validation set. Figure 5 shows the results of the training, showing how the windows extracted from the validation set are classified. It is possible to notice that the major difficulty for the network is to recognize in a correct way when the wobble board is used with movements that do not match the four proposed ones.

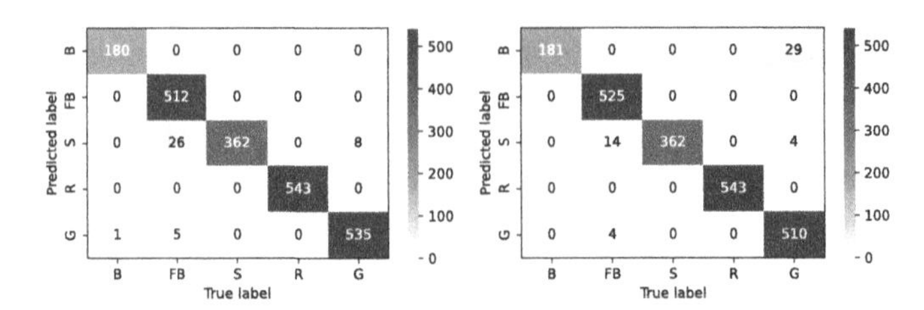

Fig. 5. Validation set confusion matrix, to the left the model with floating point weights and to the right fixed point weights.

5.2 Power Consumption

We measured the power consumption for each OM, for this purpose the digital oscilloscope ANALOG Discovery 2 was used to measure the voltage on the shunt resistor placed in series to the power cable of the SensorTile node. Figure 6 shows the result of the measurement.

Operating mode *raw*. This is the OM with the highest amount of data to be sent via Bluetooth, the minimum system frequency to handle data traffic with the Bluetooth module present in the SensorTile module is 8 MHz. In order to optimize data sending via Bluetooth, four sensor acquisitions are merged for each packet, reducing the data sending frequency 100 Hz 25 Hz.

Operating mode *balance*. In contrast to the previous one, this is the OM where there is less data transmission, in fact, the evaluation of the exercise is done every one second, invoking Bluetooth transmission at the same frequency. It has been tested that a system frequency of 2 MHz is sufficient to meet the real-time constraints, Fig. 6 shows the savings due to dynamic optimization of the system frequency.

Operating mode *CNN*. It was chosen to send the information about the classification result of the exercise every time the neural network inference is performed. In order to correctly execute the neural network and at the same time respect the real-time constraints, a system frequency of 4 MHz has been set. The length of the input frame is obviously the same as that used during training, while the distance between frames in this evaluation phase, as for operating mode *Balance*, is one second. For this reason, the power consumption is data-dependent and the worst case will thus be taken into account for the calculation of power consumption. The maximum number of times a single data item is sent via the Bluetooth module is equal to $(60 - 15)/1 + 1 = 46$, i.e., the number of 15-second frames that can be extracted from a 60-second signal considering the frame-to-frame distance equal to one second.

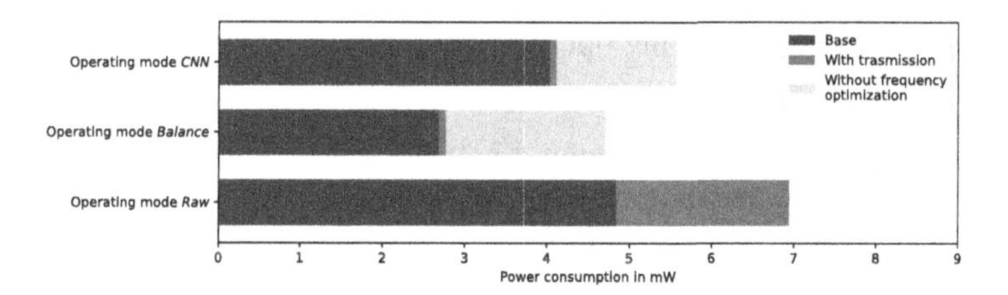

Fig. 6. Power consumption for each OMs.

5.3 Power Consumption Model

We conducted a comprehensive set of experiments measuring energy consumption in various setup conditions. The results were used to create a model that highlighted the contribution of each task to the node's energy consumption. The energy values for each task in the process network are shown in Table 4, Table 5 instead shows the power consumption of the platform as a function of the chosen system frequency.

Table 4. Summary of consumption and execution time for each task.

Task type	*Number of cycles*	*Execution time (8 MHz)*	*Energy contribution*
Get data	841	$105\,\mu s$	$E_g = 2.96\,\mu J$
Get data + balance	$1\,550 + 841$	$300\,\mu s$	$E_{gb} = 3.76\,\mu J$
CNN	2 219 582	$277\,ms$	$E_c = 852.38\,\mu J$
Threshold	910	$114\,\mu s$	$E_t = 2.73\,\mu J$
Send data	$\sim 25\,000$	$\sim 3\,ms$	$E_s = 83.96\,\mu J$

Table 5. Summary of consumption of peripherals.

Device	*Power consumption*		
	2 MHz	***4 MHz***	***8 MHz***
Platform in idle state	$2.609\,mW$	$3.101\,mW$	$4.546\,mW$

It is then possible to obtain the equations that estimate the consumption for each OM:

$$P_{raw\ data\ OM} = (E_g + \alpha E_s) \cdot f_s + P_{idle}\,, \tag{1}$$

$$P_{basic\ balance\ OM} = E_{gb} \cdot f_s + (E_t + E_s) \cdot f_b + P_{idle}\,, \tag{2}$$

$$P_{cnn\ processing\ OM} = E_{gb} \cdot f_s + (E_c + E_t + E_s) \cdot f_c + P_{idle}\,. \tag{3}$$

In Eqs 1, 2 and 3, the following operators are used:

- f_s is the sampling frequency,
- f_c frequency of convolutional neural network activation,
- f_b is the basic balance data sanding frequency,
- α^{-1} is the number of samples inserted in a BLE package,
- P_{idle} power consumption of the platform in idle state, depends on the system frequency.

Figure 7 shows graphically the contribution of each task to the power consumption of each OM.

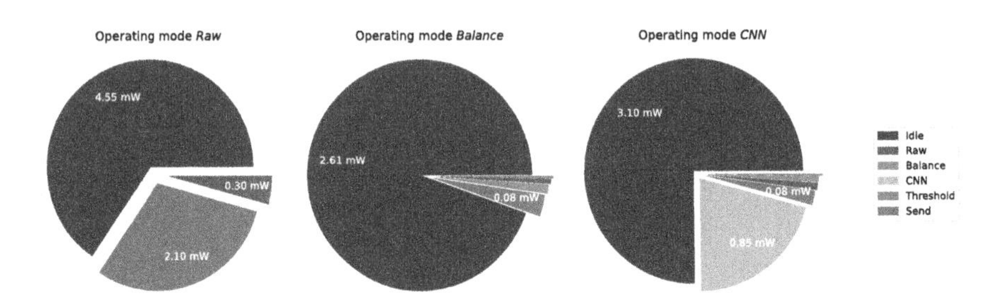

Fig. 7. Estimation of energy consumption for each task of each OM.

6 Conclusion

We defined a hardware/software template for the development of a dynamically manageable sensory node, which was addressed to perform in-place analysis of sensed data. Its implementation has been tested on a low-power platform capable of recognizing simple movements on a wobble board using CNN-based data analysis. The device can reconfigure itself based on the operating modes and workload that are required. The ADAM component, which can manage device reconfiguration, contributes significantly to energy savings. On a custom dataset, a quantized neural network achieves an accuracy value greater than 97%. By activating in-place analysis and managing the device's hardware and software components, we were able to save up to 60% on energy. This work demonstrates the feasibility of increasing battery lifetime with near-sensor processing while also emphasizing the significance of data-dependent runtime architecture management.

References

1. World Health Organization (WHO): Global recommendations on physical activity for health. https://www.who.int/dietphysicalactivity/global-PA-recs-2010.pdf. Accessed 11 Nov 2021

2. Abedtash, H., Holden, R.J.: Systematic review of the effectiveness of health-related behavioral interventions using portable activity sensing devices (PASDs). J. Am. Med. Inform. Assoc. **24**(5), 1002–1013 (2017). https://doi.org/10.1093/jamia/ocx006
3. Ghasemzadeh, H., Jafari, R.: Ultra low-power signal processing in wearable monitoring systems: a tiered screening architecture with optimal bit resolution. ACM Trans. Embed. Comput. Syst. **131**, 9:1–9:23 (2013). http://doi.acm.org/10.1145/2501626.2501636
4. Wang, C., et al.: A low power cardiovascular healthcare system with cross-layer optimization from sensing patch to cloud platform. IEEE Trans. Biomed. Circuits Syst. **13**(2), 314–329 (2019)
5. Adimulam, M.K., Srinivas, M.B.: Ultra low power programmable wireless ExG SoC design for IoT healthcare system. In: Perego, P., Rahmani, A.M., TaheriNejad, N. (eds.) MobiHealth 2017. LNICST, vol. 247, pp. 41–49. Springer, Cham (2018). https://doi.org/10.1007/978-3-319-98551-0_5
6. Tabal, K.M.R., Caluyo, F.S., Ibarra, J.B.G.: Microcontroller-implemented artificial neural network for electrooculography-based wearable drowsiness detection system. In: Sulaiman, H.A., Othman, M.A., Othman, M.F.I., Rahim, Y.A., Pee, N.C. (eds.) Advanced Computer and Communication Engineering Technology. LNEE, vol. 362, pp. 461–472. Springer, Cham (2016). https://doi.org/10.1007/978-3-319-24584-3_39
7. Magno, M., Pritz, M., Mayer, P., Benini, L.: DeepEmote: towards multi-layer neural networks in a low power wearable multi-sensors bracelet. In: 7th IEEE International Workshop on Advances in Sensors and Interfaces (IWASI), pp. 32–37 (2017)
8. Scrugli, M.A., Loi, D., Raffo, L., Meloni, P.: A runtime-adaptive cognitive IoT node for healthcare monitoring. In: Proceedings of the 16th ACM International Conference on Computing Frontiers, ser. CF 2019. Association for Computing Machinery, pp. 350–357 (2019). https://doi.org/10.1145/3310273.3323160
9. Nilsson, N.C., Serafin, S., Nordahl, R.: Gameplay as a source of intrinsic motivation for individuals in need of ankle training or rehabilitation. Presence **21**(1), 69–84 (2012)
10. Blažica, B., Krivec, P.: Olok boardy - gamified sensorimotor training with affordable smart balance board. In: 3rd Annual Scientific and Professional International Conference "Health of Children and Adolescent", September 2019, p. 185 (2019). https://www.hippocampus.si/ISBN/978-961-7055-73-3.pdf
11. Maclean, N., Pound, P.: A critical review of the concept of patient motivation in the literature on physical rehabilitation. Soc. Sci. Med. **50**(4), 495–506 (2000)
12. S.E. Asp, et al. (Eds.): WobbleActive (2007)
13. Goodfellow, I.J., Bengio, Y., Courville, A.C.: Deep Learning, ser. Adaptive Computation and Machine Learning series. MIT Press (2016). https://books.google.it/books?id=Np9SDQAAQBAJ

Towards Real-Time and Energy Efficient Siamese Tracking – A Hardware-Software Approach

Dominika Przewlocka-Rus(✉) and Tomasz Kryjak

Embedded Vision Systems Group, Computer Vision Laboratory,
Department of Automatic Control and Robotics, AGH University of Science and Technology, Krakow, Poland
{dominika.przewlocka,tomasz.kryjak}@agh.edu.pl

Abstract. Siamese trackers have been among the state-of-the-art solutions in each Visual Object Tracking (VOT) challenge over the past few years. However, with great accuracy comes great computational complexity: to achieve real-time processing, these trackers have to be massively parallelised and are usually run on high-end GPUs. Easy to implement, this approach is energy consuming, and thus cannot be used in many low-power applications. To overcome this, one can use energy-efficient embedded devices, such as heterogeneous platforms joining the ARM processor system with programmable logic (FPGA). In this work, we propose a hardware-software implementation of the well-known fully connected Siamese tracker (SiamFC). We have developed a quantised Siamese network for the FINN accelerator, using algorithm-accelerator co-design, and performed design space exploration to achieve the best efficiency-to-power ratio (determined by FPS and used resources). For our network, running in the programmable logic part of the Zynq UltraScale+ MPSoC ZCU104, we achieved the processing of almost 50 frames-per-second with tracker accuracy on par with its floating point counterpart, as well as the original SiamFC network. The complete tracking system, implemented in ARM with the network accelerated on FPGA, achieves up to 17 fps. These results bring us towards bridging the gap between the highly accurate but power-demanding algorithms and energy-efficient solutions ready to be used in low-power, edge systems.

Keywords: Siamese tracker · quantised neural networks · hardware-software implementation · energy efficient tracking · real time tracking

1 Introduction

Visual object tracking is a component of many different advanced computer vision systems, used, among others, in surveillance systems, advanced driver assistance systems (ADAS), or autonomous vehicles, such as cars or drones. Due to the high complexity of the considered problem – the tracked object can undergo changes,

K. Desnos and S. Pertuz (Eds.): DASIP 2022, LNCS 13425, pp. 162–173, 2022.
https://doi.org/10.1007/978-3-031-12748-9_13

such as rotations, occlusions, or different scene illuminations – it is still a research area of very high activity, well documented each year by the Visual Object Tracking Challenge (or VOT Challenge). Tracking methods can be roughly divided into classic (mean-shift, CAM-shift, KLT) and AI-based ones (including correlation filters). It is the development of deep learning that has allowed for significant progress in the field of tracking, and nowadays top trackers are based on neural networks, including the Siamese neural networks. Unfortunately, state-of-the-art trackers are usually characterised with a very high computational complexity (resulting, inter alia, from the very fact of using neural networks) and to achieve real-time processing, these algorithms are accelerated using high-end and energy-inefficient GPUs. At the same time, in many real-life applications, the real-time and energy-efficient processing constraints have to be met while ensuring high-quality tracking. One of the possible solutions is the acceleration of state-of-the-art trackers using SoC FPGA (System on Chip Field Programmable Gate Arrays) platforms, which allow for high parallelisation of computations, with low energy consumption. Nevertheless, this choice results in other challenges, mainly due to the limited number of resources in FPGA devices. In view of the above, in this work we propose a fast and energy-efficient hardware-software implementation of the SiamFC tracker [1], achieving accuracy on par with that of its original counterpart. The main contributions of this paper are:

- The hardware-software implementation of a Siamese tracker on the Zynq UltraScale+ MPSoC ZCU104 platform, with a detailed time analysis of the algorithm components and a design space exploration showing the relation between the power used (used resources) and the achieved speed (measured in FPS).
- The proposed algorithm-accelerator co-designed architecture of a Siamese neural network, which resulted in tracking accuracy on par with the original SiamFC approach, while significantly reducing the number of parameters (thus calculations).

To the authors' best knowledge, this is the first paper to give such a comprehensive analysis of hardware-software implementation of the SiamFC tracker.

The remainder of this paper is organised as follows. In Sect. 2 we briefly describe the concept of Siamese trackers and discuss related work. The proposed quantised Siamese tracker and its hardware-software implementation are presented in Sect. 3. The proposed tracker is evaluated both in software and hardware, with extensive accuracy, time and power consumption analysis. The obtained results are discussed in Sect. 4. The paper ends with conclusion and future research proposals.

2 Siamese Tracking

A Siamese network is a Y-shaped network with two branches joined in one output. It measures the similarity of the two inputs, thus it can be considered as a similarity function. Many of the Siamese-based trackers rely solely on this assumption. In general, we examine the following two inputs: the exemplar image of an object (from the first frame) and the region of interest (ROI), where we presume that the target is present in the following frames. Each branch processes one image, and their outputs are joined using correlation. This results in a similarity map (or maps) between object features and ROI, based on which the target can be located. Over the past years, this basic version of the Siamese tracker has undergone many modifications, affecting both the tracking efficiency and frame processing time; the most recognisable are described in Sect. 2.1.

2.1 Related Work

Fully convolutional Siamese trackers were first introduced in the paper [1]. Both the object and ROI are processed by identical branches, based on the well-known AlexNet DCNN (Deep Convolutional Neural Network) architecture. The feature maps obtained are cross-correlated to produce a single heat map, determining the location of the target centre. The ROI is analysed in multiple scales – the one for which the heat map has the highest peak is chosen to rescale the previous bounding box. A direct continuation of the research is presented in [2], where the previous solution was extended with a correlation filter as an additional layer of the Siamese network. To overcome the issue of too deep Siamese networks, in [7], a dense block-based network architecture was proposed. Each dense block is built of multiple convolution layers, the outputs of which are feed-forwarded to all next blocks. In this way, both low-level and high-level features are cross-correlated, which enhances the network's generalisation ability. In addition, the ROI branch was equipped with an attention module.

To avoid multiscale search, in [6] the Siamese-RPN framework was proposed. It consists of the Siamese network for features' extraction (ended with a cross-correlation) and the two-branch Region Proposal Network: one for foreground-background classification and the other for proposal refinement. In [5] the extension of SiamRPN was proposed. The output of the Siamese network is extended to aggregate the outputs of the intermediate layers. This allows the similarity map to be calculated using features learnt on multiple levels. Moreover, the correlation layer is replaced with a depth-wise separable correlation, which results in a multichannel similarity map with different semantic meanings for each channel. Also in this work, the authors proposed a different backbone than previously used [1,2,6] – instead of AlexNet modification, they developed an appropriately adjusted ResNet50 and pointed out the conditions to be met by deep networks to be used in Siamese trackers.

In [9], the authors proposed a VGG-16 like backbone and indicated that the most commonly used AlexNet has limited feature extraction capabilities. Unlike [5,6] based on anchors, as well as [1,2] with multiscale search, the solution proposed in [4] reformulated the tracking problem as a joint regression and classification task. Depth-wise correlation is applied to aggregated Siamese network output (with outputs from intermediate layers), and then the result is passed to two networks: one for foreground-background classification and the other for bounding box regression. In [8], benefiting from depth-wise cross-correlation, the authors proposed a new approach with the network output representing the binary segmentation mask for the target. This also enabled the prediction of rotated bounding boxes.

The works listed above do not exhaust the progress that is constantly being made in the field of Siamese-based trackers, but serve more as an overview of the most important concepts: different backbones, dealing with scale changes with multiscale search, anchors or bounding box regression, and finally the analysis of output features of multiple levels. A summary of existing algorithms with a description of current trends can be found in [3].

The listed algorithms are characterised with high tracking accuracy; however, to achieve real-time processing, they are run on high-end GPUs. There are only a few works on the low-power and real-time implementation of Siamese trackers. In [11] the authors presented preliminary results on the optimisation of fully connected Siamese networks for object tracking. With various experiments on quantisation and backbone architecture, they showed that precision reduction can positively affect the overfitting, thus also tracking accuracy. Similarly in [12] the authors focused on optimisation of the size of the Siamese network's architecture which, however, significantly influenced the effectiveness of the tracker. On the other hand, in the paper [13], the results on effective co-design for algorithm and accelerator for AI on edge were presented, also for networks typically used in Siamese trackers. In [10] the authors proposed a hardware-software implementation of a SiamRPN-like tracker in PYNQ (ZCU 104). The PS (Processor System) part is used for system configuration, reading the input frames, communication with accelerator, and displaying the results. In PL (Programmable Logic) two networks are accelerated: Siamese and Region Proposal. The authors report that their tracker runs with 36.7 FPS. Unfortunately, in the cited paper, there is no information on resource usage or energy consumption, as well as it is not clearly stated what network architecture was used and if (and how) it was quantised, which both have a direct impact on hardware implementation feasibility. Moreover, the tracker accuracy is not provided, nor is the comparison with the baseline (software) solution. Similarly in [14] the authors proposed the hardware implementation of a lightweight Siamese network using both pruning and quantisation. The tracker, running on ZedBoard with a ZCU 102 core, achieves 18.6 FPS. However, since the article lacks a description of other than network accelerator components, that is, acquisition of input data or post-processing of network's output to obtain the location of the target, it seems that the FPS rate refers solely to the neural network (not the complete tracking system).

On the basis of this analysis, one can notice a significant progress in the Siamese tracker domain (measured mainly by trackers' accuracy), which, however, is not accompanied by equally extensive research on their embedded devices' deployment. The few existing works on hardware acceleration of Siamese trackers lack of important details which makes them hard to compare. At the same time, since in many applications we face the challenge of real-time and energy-efficient processing, choosing such hardware may be necessary. This directly motivates our research.

Table 1. The proposed network architecture

Layer	Kernel	Filters No.	Quantisation	Maxpooling
Conv 1.1	3×3	64	8 bits	2×2
Conv 1.2	3×3	64	4 bits	2×2
Conv 2	3×3	128	4 bits	2×2
Conv 3	3×3	128	4 bits	-
Conv 4	3×3	128	4 bits	-
Conv 5	3×3	128	8 bits	-

3 Quantised Siamese Tracker

The most commonly used backbones for Siamese trackers are appropriate modifications of AlexNet or ResNet networks. Nevertheless in case of embedded devices implementations, one of the key elements in network architecture selection is most of all the accelerator design and its limitations (resulting also from the limited on-board resources). In this work, for network acceleration, we use the FINN framework [18,19] and the Zynq UltraScale+ MPSoC ZCU104 platform. For algorithm-accelerator co-design, in particular, one has to take into account the following factors:

- Computations precision, which results directly from the number of bits for the coding of the weights and activations. Apart from the reduction of needed memory, this also affects the number of resources used for arithmetical operations (e.g. floating point operations are far more complex than 8- or even 4-bit integer ones).
- Unified and small filters positively affect the possibility of computations parallelisation.
- Using a too deep network architecture can negatively affect the possibility of parallelisation (or, in the extreme case, cannot be implemented on a given platform).
- Custom and specific architectures may not be supported for the chosen FINN accelerator.
- Careful tuning of the folding parameters can increase parallelisation, thus decrease processing time, but at the cost of the number of resources used.

Given the above, we have designed a custom Siamese network architecture presented in Table 1 (one branch) – for all layers, we used zero padding and stride equal to 1. After each convolution layer, except for the last, there is a batch normalisation layer. The input image, the ROI, is of size $238 \times 238 \times 3$, while the one representing the object to track (used for initialisation) is of size $110 \times 110 \times 3$. The activations are quantised to 4 bits, while using 8 bits precision for weights of the first and last layers allows to maintain high accuracy.

The proposed tracking algorithm is based on SiamFC [1], which does not use the aggregation of outputs from intermediate layers, additional branches for

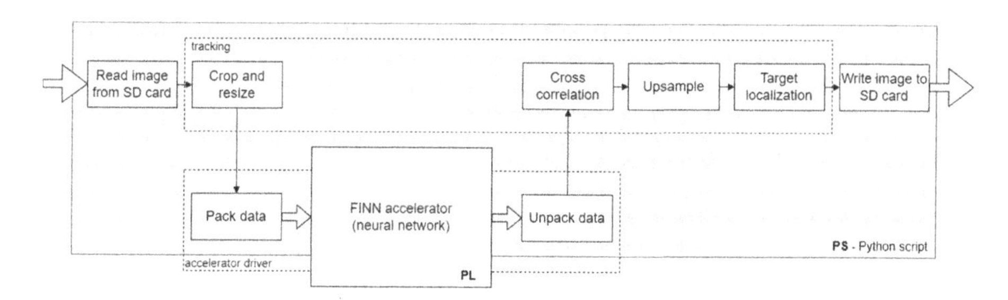

Fig. 1. Overview of the proposed hardware-software system. A single branch of the Siamese network is accelerated using the FINN framework in PL (FPGA). The Python script is run on the ARM processor (PS), handling the input and output, communicating with the accelerator and post-processing the network output.

classification or bounding box regression and other complex elements (for full algorithm description please refer to the original work). It is especially important given the choice of a FINN accelerator, which does not support most of these operations straightforwardly. Obviously, this constrains the possibility of accelerating the best existing tracker and will be widely commented on in Sects. 4 and 5. Still, FINN allows to adapt the accelerator architecture to the chosen network (unlike e.g. Vitis AI) – based on the properly prepared network graph and folding parameters, the hardware (accelerator) is generated. Folding parameters control the level of computations parallelisation: for each layer, we can set the number of simultaneously processed input channels (PE parameter) and the number of aggregated output channels (SIMD).

3.1 Hardware-Software Implementation

In this paper, we use the hardware-software approach and divide the implementation of the tracker into the network accelerated in PL and the rest of the tracking algorithm implemented in PS. The Python script is run in ARM, which is responsible for: (1) input and output handling; (2) communication with FPGA via a proper driver; (3) realisation of the tracker's logic – cropping and scaling the input image, and then post-processing the output of the network (determining the target location based on similarity map). An overview of the proposed system is presented in Fig. 1. The software part of our tracker (in an ARM processor) is run with Python 3 interpreter, using numpy, PyTorch, and OpenCV libraries. The clock for the PL is set to 100 MHz.

4 Results

We have evaluated the proposed solution in two ways. Firstly, we have tested the developed quantised tracker on different datasets and compared its accuracy with the baseline model, as well as the original solution. Second, we have done a design space exploration with different hardware settings to analyse the tracker performance. The details of these experiments are summarised below.

Table 2. Comparison of the performance of the tracker. The results were obtained using the GOT 10k toolkit. The mean average overlap (mAO) metric takes into account the potential class imbalance in the evaluation by updating the standard AO (denoting the average overlaps between all ground-truth and estimated bounding boxes) with weights proportional to the number of frames in each sequence [15] (s – scale)

Tracker	VOT 2016
	mAO
FP32 3s	0.362
FP32 1s	0.315
Quantised 3s	0.355
Quantised 1s	0.281
Original SiamFC [1] (3s)	0.385 †

† raw results downloaded from official VOT2016 challenge [17]

4.1 Benchmark Results

To properly evaluate the proposed tracker, we have prepared two versions of the network described in Sect. 3: floating point baseline and quantised. Both networks were trained in the GOT 10k dataset [15] (unlike the original SiamFC tracker, trained on ImageNet), for 50 epochs, with an initial learning rate 1e−2, reduced each epoch to a final value of 1e−5. Next, we conducted multiple experiments to compare the tracker accuracy for different scenarios: floating-point network, quantised network, and processing of single or three scales. The tracker was evaluated on the VOT 2016 dataset for a proper comparison with the original SiamFC [1].

Table 2 summarises the obtained results (we do not present the comparison with other Siamese tracker FPGA accelerators since in previous works – summarised in Sect. 2.1 – authors either do not report any accuracy results, or use other metrics): (1) For the VOT 2016 benchmark with the proposed network, our tracker achieves accuracy on par with the original SiamFC [1] when processing 3 scales regardless of the quantisation: the FP32 3s tracker is around 6% behind the original, while the quantised 3s around 8%. However, it is crucial to notice that the proposed network is far more compact than the AlexNet-based one. Specifically, the AlexNet backbone has 3747200 parameters, while ours 554688, which is around $6.7x$ less. (2) For our tracker, the best accuracy (measured by mAO) is obtained using the FP32 3 scale network. Nevertheless, after quantisation we observe only a slight decrease in accuracy - from 0.362 to 0.355 (less than 2%). (3) The decrease in accuracy is greater after reducing the number of processed scales from 3 to 1. For the FP32 network, the difference is around 0.032 (9%), while for the quantised one, even 0.074 (around 21%). Figure 2 shows an exemplar output of the quantised 1 scale tracker.

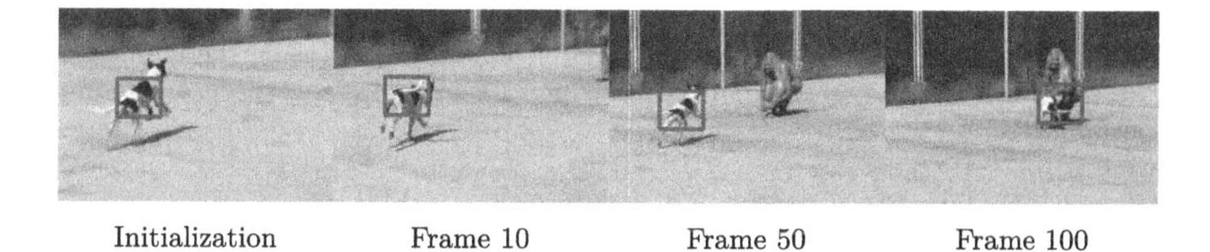

Fig. 2. Output from the quantised Siamese 1 scale tracker for 'Dog' sequence from OTB.

Table 3. Different folding settings for the FINN accelerator

	Layers' (PE, SIMD)					
	1	2	3	4	5	6
V1	(32, 3)	(32, 16)	(16, 16)	(8, 16)	(8, 16)	(8, 8)
V2	(32, 3)	(32, 16)	(16, 16)	(8, 16)	(8, 16)	(16, 8)
V3	(32, 3)	(32, 16)	(16, 16)	(16, 16)	(16, 16)	(16, 8)
V4	(32, 3)	(32, 16)	(16, 16)	(16, 16)	(16, 16)	(16, 16)
V5	(32, 3)	(32, 16)	(32, 16)	(32, 16)	(32, 16)	(32, 16)
V6	(32, 3)	(32, 16)	(32, 16)	(32, 32)	(32, 32)	(32, 32)

4.2 Performance

To obtain the best network acceleration performance using FINN we performed a design space exploration for choosing the right folding parameters (see Sect. 3). The set parameters for six different experiments are summarised in the Table 3. After hardware generation we analysed the used resources, power consumption, and latency of the network input processing. The power consumption was estimated using Vivado tools. The results are summarised in the Table 4.

For experiments V1, V2, V3 and V4 we gradually increase the number of, first, PE elements, and then SIMD, which results in a slight decrease in the used resources: mainly LUTs, responsible for arithmetical operations, but also BRAMs. At the same time, we observe a stable but subtle increase in FPS, from 38 to 42, with a simultaneous increase in power consumption of around 0.5 W. A considerable change was achieved after doubling the number of PEs in layers 3, 4, 5, 6 – for experiment V5, in relation to V4. The number of LUTs used increased by around 18%, BRAMs 5%, FFs 5%, which accelerated processing by around 7 FPS to 49 FPS, with an increase in power consumption of 0.6 W. Interestingly, next experiments with increasing the level of parallelisation – V6, where we double the number of SIMDs for layers 4, 5, 6 – caused considerable increase in the used resources (over 90% available LUTs and BRAMs) and the power consumption to almost 7W, while improving processing speed by only 0.6 FPS. The dependence between power consumption and the FPS achieved is presented in Fig. 3.

Table 4. Comparison of accelerated Siamese network performance for different folding configurations. When increasing the level of parallelisation (using the number of PEs and SIMDs), we can observe both an increase in processing speed and power consumption

Folding	Resources				FPS	Power [W]
	LUT	FF	BRAM	LUTRAM		
V1	40.45%	16.78%	46.31%	11.92%	38.46	4.5
V2	42.25%	17.44%	50.8%	12.1%	40.24	4.56
V3	46.66%	17.9%	50.8%	12.1%	41.31	4.81
V4	48.72%	18.6%	50.8%	12.16%	42.16	4.92
V5	66.87%	23%	55.29%	12.54%	49.03	5.5
V6	91.27%	28.66%	91.83%	13.58%	49.63	6.79

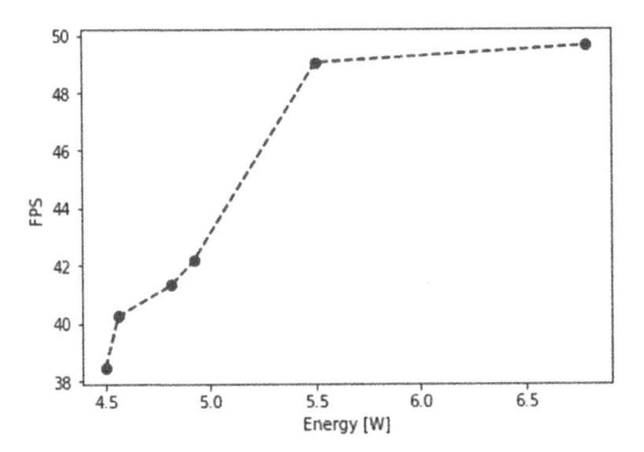

Fig. 3. Design space exploration for acceleration of Siamese network with power to FPS ratio. The used power is tightly connected to the used resources presented in Table 4.

Table 4 shows the results only for network acceleration and does not take into account transfers to and from accelerator, as well as the rest of the tracker computations. Careful time analysis for the complete software-hardware implementation, using the V5 accelerator, is presented in Table 5, from which we draw several conclusions. (1) The network acceleration takes 50% of the time needed to process the ROI. This also includes the time for packing the input data, transferring them to the accelerator, from accelerator, and unpacking. In other words, acceleration with data transfer enables the processing of a single ROI with a speed of around 35 FPS. (2) Almost half the time for post-processing of the network output (around 14% of total) is used for cross-correlation, while around 10% for target location (with, among others, cosine window filtration). (3) The network input pre-processing (cropping and scaling ROI) takes relatively little time. Much greater impact on the processing speed is the transfer of data from the accelerator (and unpacking) than the transfer to the accelerator (including packing) – 14% vs 2% of total time. Finally, the complete hardware-software tracking system processes a frame with around 17 FPS.

Table 5. Analysis of the average latency of each tracking stage for the V5 folding version. The network acceleration with I/O data transfers achieves around 35 FPS, while the complete tracking system operates at a speed of 17 FPS.

Stage	Time [ms]
Crop & resize	0.0102
Input transfer	0.001
Network acceleration	0.0205
Output transfer	0.008
Cross correlation	0.0081
Upsampling	0.0011
Locating target	0.0057
Sum	**0.0546**
Total (measured)*	0.0587
Input preprocessing	0.0102 (18%)
FINN network transfer & execution	0.0295 (52%)
Network output processing	0.0149 (25%)

*with other additional operations

5 Conclusion

In this work, we have proposed a hardware-software implementation of a Siamese tracker, based on [1]. Firstly, we have designed a Siamese neural network, which architecture meets the chosen FINN accelerator constraints, and at the same time allows our tracker to achieve the accuracy on par with the original SiamFC solution, even for the quantised version. Second, we have performed a design space exploration, increasing the level of paralellisation in FINN accelerator and have shown the relation between the power consumption and tracker speed. Finally, our tracker achieves around 17 FPS with 5.5 W power consumption. The original tracker run on NVIDIA GeForce GTX Titan X with 250 W power consumption, achieved 83 FPS [1]. We have also provided a time analysis of each tracker component and pointed out the bottlenecks of the proposed solution. On the basis of that, we draw two main conclusions for future work:

- Despite the fact that our tracker is on par with the original SiamFC, the accuracy achieved is far behind the best existing Siamese tracking algorithms. The next work should then be supplemented with different SoTA features, such as bounding box regression or aggregation of features from different levels. It is important to note that acceleration of such a network would not be possible using FINN in a straightforward manner. Therefore, for future work, it is planned to usethe FINN accelerator as one of the hardware components, along with some other, custom solution for e.g. bounding box regression.

- In the current version of the tracker, data transfer from the accelerator and output analysis have a big impact on the latency of the solution (almost 40% of total time). Moving the post-processing to FPGA would significantly improve the frame processing time, since the output transfer to PS would not be needed, and, at the same time, the cross correlation could be parallelised.

Based on the above, we also want to pay attention to the fact that the progress in developing more and more accurate tracking algorithms (including the Siamese-based ones) is far beyond the progress in AI on edge deployment, especially for available, ready-to-use accelerators. Such solutions usually do not support various advanced methods standard for software approaches, at least not without a deep interference in the source code (which is still feasible only for rare open-source solutions). Faced with the need to significantly reduce the energy demand, both for deployment on low-power devices and for global needs, we believe that continuous work on energy-efficient advanced vision systems is especially important.

Acknowledgements. The work presented in this paper was supported by the National Science Centre project no. 2016/23/D/ST6/01389 entitled "The development of computing resources organisation in latest generation of heterogeneous reconfigurable devices enabling real-time processing of UHD/4K video stream". The authors would like to thank Joanna Stanisz and Konrad Lis for they supprort when working with FINN, in particular on the reduction of the data transfer time to and from the accelerator.

References

1. Bertinetto, L., Valmadre, J., Henriques, J.F., Vedaldi, A., Torr, P.H.S.: Fully-convolutional Siamese networks for object tracking. In: Hua, G., Jégou, H. (eds.) ECCV 2016. LNCS, vol. 9914, pp. 850–865. Springer, Cham (2016). https://doi.org/10.1007/978-3-319-48881-3_56
2. Valmadre, J., Bertinetto, L., Henriques, J., Vedaldi, A., Torr, P.: End-to-end representation learning for correlation filter based tracking. In: Proceedings of the IEEE Conference on Computer Vision and Pattern Recognition (CVPR) (2017)
3. Ondrašovič, M., Tarábek, P.: Siamese visual object tracking: a survey. IEEE Access **9**, 110149–110172 (2021)
4. Guo, D., Wang, J., Cui, Y., Wang, Z., Chen, S.: SiamCAR: Siamese fully convolutional classification and regression for visual tracking. CoRR. abs/1911.07241 (2019). http://arxiv.org/abs/1911.07241
5. Li, B., Wu, W., Wang, Q., Zhang, F., Xing, J., Yan, J.: SiamRPN++: evolution of Siamese visual tracking with very deep networks. CoRR. abs/1812.11703 (2018). http://arxiv.org/abs/1812.11703
6. Li, B., Yan, J., Wu, W., Zhu, Z., Hu, X.: High performance visual tracking with Siamese region proposal network. In: 2018 IEEE/CVF Conference on Computer Vision and Pattern Recognition, pp. 8971–8980 (2018)

7. Abdelpakey, M.H., Shehata, M.S., Mohamed, M.M.: DensSiam: end-to-end Densely-Siamese network with self-attention model for object tracking. In: Bebis, G., Boyle, R., Parvin, B., Koracin, D., Turek, M., Ramalingam, S., Xu, K., Lin, S., Alsallakh, B., Yang, J., Cuervo, E., Ventura, J. (eds.) ISVC 2018. LNCS, vol. 11241, pp. 463–473. Springer, Cham (2018). https://doi.org/10.1007/978-3-030-03801-4_41
8. Wang, Q., Zhang, L., Bertinetto, L., Hu, W., Torr, P.: Fast online object tracking and segmentation: a unifying approach. CoRR. abs/1812.05050 (2018). http://arxiv.org/abs/1812.05050
9. Li, Y., Zhang, X.: SiamVGG: visual tracking using deeper Siamese networks (2019)
10. Cui, Z., An, J.: Heterogeneous Siamese tracking system based on PYNQ framework. In: 2020 6th International Conference On Control, Automation And Robotics (ICCAR), pp. 16–20 (2020)
11. Przewlocka, D., Wasala, M., Szolc, H., Blachut, K., Kryjak, T.: Optimisation of a Siamese neural network for real-time energy efficient object tracking. In: Chmielewski, L.J., Kozera, R., Orłowski, A. (eds.) ICCVG 2020. LNCS, vol. 12334, pp. 151–163. Springer, Cham (2020). https://doi.org/10.1007/978-3-030-59006-2_14
12. Cao, Y., Ji, H., Zhang, W., Shirani, S.: Extremely tiny Siamese networks with multi-level fusions for visual object tracking. In: 2019 22th International Conference on Information Fusion (FUSION), pp. 1–7 (2019)
13. Hao, C., et al.: Effective algorithm-accelerator co-design for AI solutions on edge devices. (2020). https://arxiv.org/abs/2010.07185
14. Zhang, B., Li, X., Han, J., Zeng, X.: MiniTracker: a lightweight CNN-based system for visual object tracking on embedded device. In: 2018 IEEE 23rd International Conference On Digital Signal Processing (DSP), pp. 1–5 (2018)
15. Huang, L., Zhao, X., Huang, K.: GOT-10k: a large high-diversity benchmark for generic object tracking in the wild. IEEE Trans. Pattern Anal. Mach. Intell. **43**, 1562–1577 (2021)
16. GOT 10k leaderboard. http://got-10k.aitestunion.com/leaderboard. Accessed 28 Mar 2022
17. Hua, G., Jégou, H. (eds.): ECCV 2016. LNCS, vol. 9914. Springer, Cham (2016). https://doi.org/10.1007/978-3-319-48881-3
18. Blott, M., et al.: FINN-R: an end-to-end deep-learning framework for fast exploration of quantized neural networks. ACM Trans. Reconfigurable Technol. Syst. **11**, 1–23 (2018)
19. Umuroglu, Y., et al.: FINN: a framework for fast, scalable binarized neural network inference. In: Proceedings of the 2017 ACM/SIGDA International Symposium on Field-Programmable Gate Arrays, pp. 65–74 (2017)

Author Index

Printed in the United States
by Baker & Taylor Publisher Services